Assessment in a Multicultural Society

BIOLOGY AT 16+

a discussion document

Michael Vance

LONGMAN

1984

This title, an outcome of work done for the Schools Council before its closure, is published under the aegis of the School Curriculum Development Committee, Newcombe House, 45 Notting Hill Gate, London W11 3JB.

Published by Longman
Longman Resources Unit
33-35 Tanner Row
York YO1 1JP

CONTENTS

FOREWORD

In this introductory note to this series of documents I would like to make it clear that these are discussion documents which the Schools Council thinks it helpful to publish, and that they are not statements of Schools Council policy. The views expressed in them are those of their authors, and not necessarily the views of the Schools Council.

It is part of the business of the Schools Council to stimulate discussion and ideas within the field of curriculum and assessment; where there is no discussion and no disagreement there is no life. Some of the ideas and views expressed in these documents may be regarded as critical or controversial, but it is the Council's view that discussion of the curriculum and assessment is helpful and leads either to improvements or to a better understanding of what exists in relation to needs and to possible alternatives.

These documents are therefore published in the belief that they may stimulate such discussion, which may or may not lead to changes in assessment, but should lead to a better understanding of the problems involved.

The Schools Council would like to express its thanks to the many people and bodies who have given help with the preparation of this series, and, in particular, to the examining boards for their generous cooperation which included their officers' discussions with the authors, as well as the provision of, and permission to quote from, syllabuses, papers, and examiners' reports.

ARNOLD JENNINGS

Chairman, Examinations Committee
Acting Chairman, Schools Council

INTRODUCTION

Assessment in a Multicultural Society is one of several major projects in the Schools Council's fifth programme of work, Improving the Examinations System. On 25 September 1981 a Schools Council conference on examining in a multicultural society[1] brought together representatives of the GCE and CSE examining boards, teachers, members of Her Majesty's Inspectorate and others actively involved in multicultural curriculum and examination development to discuss the relationship between examining and the needs of a multicultural society. Following the conference, a number of subject specialists were commissioned to write detailed reviews of 16+ examination provision in their subject areas, with particular reference to the needs of a multicultural society. An important consideration governing the timing of the commissions was the concurrent preparation of general and subject specific national criteria for syllabuses, forms of assessment and grading for the proposed single system of examining at 16+. Though, at the time of writing, a final decision on the single system has yet to be made, it seems clear that the national criteria, whether applied to a dual or single system, will play a crucial part in determining the future form that 16+ examining takes.

Both the conference and the ensuing project are part of a much wider series of activities that the Council has undertaken to encourage developments in multicultural education. The Council, for example, has endeavoured to ensure that the 16+ national criteria take into account the multicultural dimension. Other current work concerning the multicultural education of all pupils includes the writing of pamphlets for teachers on resources and on school and classroom practice, the development of materials to enhance sensitivity to cultural diversity and an exploration of models of in-service training of teachers. These activities complement the Council's continued work on the particular needs of children from minority groups, which includes a major project on supporting community languages in primary schools and work on the language development of pupils whose main language or dialect is not standard English.*

Cultural differences which have a religious, linguistic or ethnic basis are potentially a source of considerable conflict, and therefore a matter for serious social, political and educational concern. The Council's programme of work in multicultural education aims to reduce prejudice and hostility towards minority groups, and also to help teachers provide a positive and non-racist learning environment for all their pupils.

* This term is used in the sense defined by V. Edwards in her book *Language Variation in the Multicultural Classroom:* 'standard English [is] merely another dialect with high social acceptance and status.'

The Council believes that the development of a multicultural curriculum which includes examples drawn from a wide range of cultures and which avoids presenting a solely Anglocentric view of the rest of the world, can add new opportunities and dimensions to all children's learning. The report[2] in 1981 of an earlier Schools Council project, *Education for a Multiracial Society*, stressed the need to permeate all school subjects with a multicultural perspective, and suggested the following five curriculum guidelines which have been reaffirmed in later Council publications:

1 The curriculum needs to be both international in its choice of content and global in its perspective.

2 Contemporary British society contains a variety of cultural groups. This variety should be made evident in the visuals, stories and information offered to children.

3 Pupils should have access to accurate information about differences and similarities between cultural groups.

4 People from British minority groups and from other cultures overseas should be presented as individuals with every variety of human quality and attribute. Stereotypes of minority groups in Britain and of cultures overseas, whether expressed in terms of human characteristics, life styles, social roles or occupational status, are unacceptable and likely to be damaging.

5 Other cultures and nations have their own values; whenever possible they should be allowed to speak for themselves so that they can be better understood, and not judged in terms of British or European norms.

It is important that syllabuses and examinations should harmonize with this developing multicultural curriculum. In its evidence to the Government's Committee of Inquiry into the Education of Children from Ethnic Minority Groups,[3] the Council outlined its position with regard to examining within a multicultural society as follows:

> Public examination results are crucial to a majority of pupils. They qualify pupils for admission to higher and further education; they can determine whether or not a candidate will be considered for a particular job. It is therefore essential that all pupils have equal opportunities for demonstrating their qualities and abilities through the public examination system.
>
> The interaction between the curriculum, both 'open' and 'hidden', and examinations is very complex. Curriculum development for the fourth and fifth years of secondary education appears to be slower than for preceding years; in Schools Council Pamphlet 18, *Multi-ethnic Education: the Way Forward*, Alan Little and Richard Willey write that 'A number of authorities and 40 per cent of Headteachers, 50 per cent of those in schools with a concentration of minority ethnic group pupils, reported that examinations restrict curriculum development to reflect a multi-ethnic society.' However, the current preparation for a single system of examining at 16+, based for the first time on nationally agreed syllabus and assessment criteria, presents a crucial opportunity to innovate.

The Council's submission goes on to consider examination provision in a world that is increasingly interdependent and part of a society that is already culturally diverse. It recommends that examinations should be culturally fairer, more culturally diverse and include more and modified examinations with particular cultural emphases. It continues with detailed recommendations as follows:

Towards culturally fairer examinations

In so far as our society, like all societies, reflects in its examination system the dominant cultural values, this operates in all facets of our examination system, be it examination personnel, the syllabuses, the forms of assessment, or the awarding criteria.

The changing nature of British society requires the formulation of criteria relating to:

1 Syllabuses based on a culturally broad selection of knowledge, skills and attitudes.

2 Forms of assessment (for example, projects, course work) which enable teachers to encourage children to follow areas of study which are of particular interest to the children.

3 Content choices of examination questions that are sufficiently varied to enable all candidates to demonstrate their capabilities.

4 The language, particularly the technical and idiomatic, used for examination questions so as to minimize difficulties of comprehension.

5 Stimulus material that is used as part of examination papers and that illustrates a variety of perspectives.

Examiners should be sensitive to the difficulties of candidates in demonstrating abilities when English is not their first language, and marking should be sympathetic to candidates, rewarding attainment demonstrated through various cultural attitudes and responsive to the full range of acceptable language.

Towards culturally diverse examinations

Examinations, and thus the work leading to them, should be more culturally diverse, not only to be culturally fairer to ethnic minorities but also to enrich the lives of all candidates irrespective of their cultural background. Further, in so far as culturally diverse examination courses can influence pupils' attitudes, they can also play a positive, albeit minor role, in furthering international understanding and improving relationships between the various ethnic groups within Britain.

An apparent dilemma facing the examination world, if it wishes to encourage culturally diverse examinations, is the question of syllabus and examination question options. On the one hand, the existence of options allows teachers and candidates to concentrate on their predilections; on the other hand, fewer options ensures that, for example, Geography candidates could not restrict their studies to Europe, as sometimes happens, but also undertake some study of the Third World. Culturally diverse examinations should encourage breadth with opportunities for specialization in depth.

Towards examinations with particular cultural emphases

At present there are some GCE and CSE board-based examinations designed with particular cultural emphases, but current problems are:

1 They are available from only a few boards which the subject teacher may not be able to choose, as the school as a whole does not enter candidates for them.

2 There are more GCE Mode I than CSE Mode I examinations of this type, so the majority of candidates, for whom CSE is designed, are not so well served as the minority.

3 They tend to concentrate on language and literature, and neglect other cultural dimensions, for example: art, music, geography, history.

4 They are readily accessible to candidates from only a small range of cultural backgrounds.

5 They were often designed for a different clientele (for example, candidates from overseas), and have not changed sufficiently to meet new needs.

In order to improve the position, a number of school-based examinations have been developed in recent years, but they tend mainly to be CSE and to have been prepared by schools with a high proportion of ethnic minority pupils to suit the needs of those particular pupils. Thus GCE candidates or pupils in other schools are ill served. In addition, in so far as they are Mode IIIs, they are often dependent upon the continued presence of specific teachers, can be expensive and therefore subject to increasing economic pressure, and are sometimes regarded warily by potential employers.

The examining boards in adapting their procedures to facilitate changes in their examinations are urged to consider whether:

the composition of those examining reflects and meets the needs of a multicultural society

there is any mechanism for regular monitoring of syllabuses and examination papers in the light of multicultural considerations

there is any provision for consultation and for sharing resources with other bodies who have been developing multicultural perspectives.

In considering the role of examining boards in effecting change, the Council's evidence to the Swann Committee of Inquiry continues:

If our examinations are to be modified to meet the changing needs of a multicultural society then a necessary, though not sufficient, requirement is the creation of a consensus recognizing these needs.

Those wishing changes must recognize the genuine difficulties faced by examination boards. Nearly all change requires additional finance and the boards are subject to the economic constraints faced by the rest of the educational world. In addition, some of the examinations might, particularly initially, attract only small candidatures and thus not benefit from economies of scale.

The role of the boards in innovation depends to a large extent on the wishes of their constituent members, who often do not express any desire for change. New ideas and courses are often difficult to evaluate and to relate to existing standards. Examiners and moderators in specialist subjects are not easily available.

However, it should be pointed out that there is much existing good practice to build upon. Several of the examination boards, through their overseas examining, have extensive experience of providing examinations for candidates with differing cultural backgrounds. Also most of the examining boards have specific examples of subject examinations that meet some of the needs of today's multicultural society, and the Welsh Board offers its examinations through the medium of either English or Welsh. Apart from the very wide range of ethnic minority language and literature examinations available, there exist examination subjects such as Indian Music and Dance, and Caribbean, African or Asian Studies.

There are also numerous examples of change within the more traditional subject areas. In Art and Design there are examinations utilizing practical work derived from different cultures and customs; in Biology there are examinations with examples and applications from more than one country; in English Literature examinations offering a selection of American, African and Caribbean texts; in Geography and Environmental Studies, examinations reflecting the changing nature of our society; in History, examinations that give prominence to non-European History and that link local, national and world history to the reactions and concerns of the peoples of the former British Empire; in Home Economics, examinations that recognize the variety of lifestyles of the pupils, and in Religious Education, examinations that include not only the study of Christianity as perceived and practised by a variety of ethnic groups but also the study of other religious faiths.

With careful nourishment these examples can be the seeds of future change.

The commissioning of these subject reports was seen as a small part of the nourishment process. The 'commissionees', all subject experts with experience of multicultural and 16+ examination matters are all, bar one, practising school teachers. They were asked to undertake a six month, part-time, review of existing 16+ examination provision within their subject and to ascertain how far this provision met or could be developed to meet the needs of a multicultural society.

This pamphlet is therefore one of a series which attempts to explore the subject-specific implications of assessment in a multicultural society. Seven of the reports cover the major subjects of Art and Design; Biology; English; History; Home Economics; Religious Studies, and Social Sciences. These seven reports are intended to encourage the development of 16+ examinations within the more traditional subject areas that will draw from a wider range of cultures than hitherto and that are as a result less culturally biased and more empathetic to the multicultural nature of our society. An eighth report covers Urdu, Hindi, and Modern Greek and is intended to encourage the development of 16+ examinations with particular cultural emphases that will both better serve the current and future needs of ethnic minority candidates and also be more accessible to all pupils, irrespective of their cultural backgrounds.

All eight reports are seen, as stated in the Foreword, as discussion documents contributing to a continuing debate. Examining in a multi-cultural society is a complex and, at times, an emotive undertaking, and we are well aware that the views presented will not exactly mirror those of the reader, indeed we are conscious that some of the opinions expressed by the various authors do not even represent the views of our own committees. But this is to be expected, and is a healthy situation, for there are no easy solutions. It is hoped, however, that the ideas, criticisms, examples of 'good practice', suggestions, etc., presented within the series might help stimulate the debate and serve as catalysts for future examination development.

The first audience for the Assessment in a Multicultural Society series is thus the policy committees and subject panels of GCE and CSE examining boards and all those involved in the formulation of national criteria at 16+ examinations. It is hoped that the numerous subject-specific propositions within the reports, together with the more general points raised in this Introduction, may help this target group in their development work.

The examining boards, as a result of national criteria and the resulting examination provision, will clearly exert an important influence on the educational provision of all pupils within our schools. However, their potential role, though significant, will be limited; they can, and in many instances do, initiate change but, unless other sections of the educational profession, in particular practising teachers, recognize and indicate the necessity for change, the examination 'industry' is unlikely to move too far forward for fear of distancing itself from its 'customers'.

Thus a second, and in the long term a more crucial audience for the reports must be those teachers and teacher educators who are concerned to relate educational practices to the realities of today's and tomorrow's multicultural society.

Recent policy statements have affirmed the importance of preparation of all pupils for life in a society which is culturally diverse. The science curriculum and its assessment can make a positive contribution to this preparation by emphasizing cultural diversity in the origins, practice, and development of science and by ensuring that content includes examples from many cultures. Biology can also make a particularly important contribution by sensitive teaching about genetics and 'race'. In the interest of greater cultural fairness, the science curriculum and its assessment also needs to take particular account of the cultural diversity among the school population by recognizing the experiences, perspectives, and needs of the minority groups.

This pamphlet attempts to consider the implications of a multicultural society for science education, with particular reference to Biology. Michael Vance, a Biology teacher in a North London comprehensive school, discusses the content of present CSE, O-level, and joint 16+ Biology examinations in the light of the changing needs of our multicultural society and the proposals for a new 16+ examination system. He comments on areas of 'good practice' and offers some suggestions for change.

Biology has a particularly influential place in the science curriculum: many of the pupils entered for 16+ examinations will have Biology as their only science subject; other pupils are examined in general, integrated or combined science which has Biology as one component. We hope that this discussion document will help to develop Biology syllabuses which are appropriate to a multicultural society; we also hope that it will help other science teachers and examiners review their own subject in a similar way.

GEOFF BARDELL
Project Lead Officer

ALMA CRAFT
Coordinator for
Multicultural Education

January 1984

I MULTICULTURAL SCIENCE

1 The context

Science has tended to be excluded from the debate about the role of multicultural education within the school curriculum. The Rampton Report (1981)[4] commented on the need for the *whole* school curriculum to be sensitive to the different needs of its consumers. This view, supported by the Afro-Caribbean Educational Resource Project,[5] calls for whole school policies to be developed so that all curriculum subjects reflect different cultural traditions. The lack of materials with a broad cultural perspective which have been developed within school science indicates an assumption that science is universal, and therefore culturally neutral. Science may be objective but the cultural perimeter of science is fixed by the extent of our knowledge and by our choice as educators to select the examples and illustrations that we make available to our pupils. Biology has an important role to play in meeting the needs of minority ethnic pupils, as well as in the multicultural education of all children. Jeffcoate has argued that:

> The preservation of minority cultures in a fossilised form is not what multiracial schools should be doing ... instead, their aim should be to make them [the pupils] critically aware of their culture and equipped to decide for themselves how much to retain.[6]

Very often, images of minority cultures *are* preserved in this fossilized form, and the Biology curriculum can offer minority ethnic students explanations of their origins, practices and physiology. At the same time, all pupils need to learn to see beyond their own ethnicity both for their sakes and for the sake of our society.

For Biology, this implies that time and resources should be allocated to widening the scope of the present curriculum so that it extends pupils' experiences, and reduces the ignorance which gives rise to the 'scientifically based' ideas of stereotype, which can form the basis for prejudice. No other area of the curriculum can adequately deal with the the natural curiosity displayed by children about each other's different attributes; if unchallenged in both primary and secondary school, stereotyped explanations will be repeatedly passed on to succeeding generations of pupils.

Racial tolerance and understanding can only be promoted within the context of a multicultural curriculum and, like all subjects, Biology must include a multicultural dimension in the primary school, and nurture this in the early secondary and examination years.

Meanwhile pressure groups among the ethnic minority communities are claiming that teachers and examining boards are racist, and that they ensure under-achievement among black pupils by maintaining syllabuses which, because of their narrow content and exclusion of ethnic minority viewpoints, do little to motivate these minorities. They argue that science examination curricula, for example, have ignored the contribution of peoples from the Third World, and portray them as starving masses dependent on Western technology.[8,9]

2 Science and the multicultural curriculum

The importance of science education to the participation and status of the minority ethnic communities in a society cannot be overstated. In South Africa there is a great gulf between the ethnic groups that constitute a majority and the dominant white group in the society. Within the curriculum of South Africa there are sub-divisions of material and content for the syllabuses serving the different racial groups in their largely separate schools. In addition, theories of eugenics, racial variation, and the evolution of Homo sapiens occupy an important place in the Biology examination syllabus:

> The hierarchy of race occupies an important part of the senior science curriculum in most South African schools. The different aptitudes and characters of the races making up South African society are explored in an open and thorough manner ...[10]

The ideology and practice of apartheid in South African society is thus bolstered by Biology used for indoctrination and maintaining an unequal society based on racial discrimination. In British schools the Biology curriculum has an important responsibility in preparing pupils for life in a culturally diverse society.

The Association for Science Education (ASE) issued a consultative document *Alternatives for Science Education* in 1979, which initiated discussion about the social aspects of science, including multicultural education; the questions raised are pertinent to the development of a broader perspective through current practice, content, and relevancy:

> Science in spite of its humanistic traditions, has become a slave to the high technology society and is more concerned with system maintenance than social purpose. Syllabuses emphasise theory and restrict personal involvement.[11]

The personal involvement of students in science is increased when science is used to explain the occurrences of everyday life. Where personal involvement and experience do not play a major role in the learning environment within the science classroom/school laboratory, then the potential contribution of minority ethnic cultures to science is not being realized. The ASE document continues:

> Lack of opportunity to explore history and philosophy of science and to study science in its social, economic and political contexts is perhaps a major contributory factor to the popular image of science gained by young people. What they see is a subject dominated by facts and principles established outside their experience of everyday life. They study a subject isolated

from its history or context and which is also very demanding, time consuming and complex. This image results in part from the failure to locate science studies in the everyday context of life and work, and in part from the failure to integrate science into the common culture. The majority of young people fail to demystify the subject and do not see science for what it is — one of the most important cultural activities devised by man.[12]

The ASE could have added that science should be integrated into the diverse cultures which exist in contemporary Britain, and should thus be accessible to minority cultures also. The ASE document however, does criticize the teachers for failing to take opportunities to change the curriculum. If science educators were to adopt these principles, then multiculturalism could be incorporated into mainstream science syllabuses and cease to be labelled 'for special interest only'.

Townsend and Brittan, in perhaps one of the most searching studies into multicultural education,[13] asked school heads of department about the aims and content of subject syllabuses. Most respondents agreed that syllabuses should have as one of their aims the preparation of pupils for a multiracial society. However, only 45 per cent of science department heads responded affirmatively to the question compared to over 80 per cent of the heads of humanities departments. Townsend and Brittan concluded that those who answered in the negative felt their own particular discipline did not lend itself to the preparation of pupils for a multiracial society. However, they noted that comments such as 'science is international' were not uncommon among responses from heads of science departments, and they reported that an encouraging number of submissions expanded on how science syllabuses did include this aspect of 'international' science. Some heads of department drew attention to the fact that they were following examination syllabuses which did not include this element. It is important to bear in mind that this particular study was carried out in 1972 long before the questions surrounding multicultural education became of central importance to educational policy.

The suitability of syllabuses was tested by the question 'Do the syllabuses at present in use in your school (or subject) include any intended to prepare pupils for a multiracial aspect of society?' Only 21 per cent of the heads of science department replied 'Yes' to this question. Some departmental heads used terms such as 'science is universal' and 'science does not discriminate' in their written responses. Twenty heads of science departments listed topics exclusively within the scope of Biology, and included genetics, evolution, blood groups, skin colour, adaptations to environment, and food and nutrition. Some respondents said that 'immigrants' lacked vocabulary (with respect to the language register of science) or previous knowledge of science. One might ask whether the comment, 'lacked previous knowledge of science', reflects a stereotyped view of 'deficits' in the cultural patterns of minority ethnic groups?

In general, science department heads did not foresee developments which would make the discipline more appropriate to pupils living in a multiracial society; a few heads suggested changes towards a greater emphasis on human biology, physiology, heredity, and health education. This would have implications for Biology examinations and would militate against a 'colour blind approach'. As the Commission for Racial Equality has commented:

> [the idea] that the school curriculum develop a colour blind approach to preparing pupils for a multi-racial society will only yield further problems ...[14]

Regrettably, some twelve years since the Townsend and Brittan study, science still seems to lag behind other subject areas which have developed both resources and practice to prepare pupils for entry into a multicultural society. However, the picture is not as gloomy as it first appears. The fall in pupil numbers has led to the opening of 'new' schools, formed as a result of amalgamations especially in urban areas. Within this framework of reorganization, several of the more innovative inner city LEAs have used the opportunity to reconsider the multicultural curriculum with a view to developing whole-school policies. This has meant that a number of science departments have evaluated their contribution to the school curriculum and the needs of a culturally diverse school population. Willey refers to some of the approaches that can be taken up by a Biology department reviewing its provision:

> The curriculum should aim to promote an awareness of racial differences and their origins, and to explain these differences in terms of the biological principles of variation, evolution, natural selection and adaptation to environment.
>
> By avoiding racial-stereotyping and offence regarding religious or other observances/beliefs. Taking care in choice of illustrations of physiological and biochemical phenomena.
>
> By using examples from minority as well as the majority groups in explanations and descriptions of society and man's development. Where we deal with matters which affect race to take a broad multicultural view and attempt to avoid a negative approach when dealing with biological and environmental problems associated with Third World countries, and particularly the countries of origin of our own pupils.[15]

Further 'good practice' is also commented upon by HM Inspectorate:

> In four of the schools, work in science was seen which could be described as cross-cultural. An Asian teacher of science described the food he ate at home and compared its content and food value with comparable Western dishes. Another teacher outlined the science of pigmentation using different pupils in his class as a living resource. In yet another school genetics was being taught with sensitive reference to current theories about racial types and their supposed characteristics.[16]

HMI also remarked on the significance which two schools attached to the importance of language support specifically related to science, and this is developed in the next section.

Although science has been slow to take up the issue of the place of multicultural education within the mainstream curriculum, and indeed has failed to implement the kinds of changes that have taken place in the humanities with respect to the needs of a culturally-diverse school population, 'good practice' has been identified and the task is now to build this into new resources, examples, syllabuses and in-service training which can bring science, including biology, into the framework of a responsive and more relevant curriculum.

3 Language in science

Linguistic diversity is a feature of British schools. With the advent of non-English speaking parents whose children may not speak standard English, language has become an issue for curricular debate, especially in the sciences.[17] Although much research and innovation have been initiated within the humanities, only recently have ripples reached science, and in particular lower school science. Examples of this development have been the ASE LAMP materials, Insight to Science, and Science at Work. All of these courses show a consideration of vocabulary and the level of language used, and intersperse text with illustration. Language here is viewed in the context of collaborative working, and pupils are encouraged to discuss experiments while carrying them out, and to compare results.

The linguistic register used in science has a distinctive style that has its own grammatical structure and vocabulary, which have been developed for the purposes of classification, distinction, and definition. As Rosen, in Barnes,[18] reports, this involves the exact use of many complex linguistic rules particular to science. These include the use of non-countables, for example: 'Magnesium is a metal', not 'the Magnesium is a metal'; the passive voice is used extensively, but some pupils find it difficult to use, and others think it too impersonal. Written language in science is usually used to make precise records of experiments, and this transactional prose is heavily emphasized in the current Biology 16+ examinations. It is argued that this form of writing helps children to think through, both logically and sequentially, an argument or experimental record. Rosen views this kind of writing, as writing for an audience with the teacher as the assessor. The use of the active voice and expressive forms of language is usually discouraged. The different scientific subjects are also marked by linguistic boundaries where registers particular to the subject have evolved, and this problem is compounded for pupils from minority ethnic groups. Another difficulty arises where meanings are not translatable between languages. Where there is no word, there is no means of symbolic expression. Therefore the child for whom English is not the first language has several obstacles to overcome in the mastery of the precise and institutionalized language of the science register.

If examinations are to take into account the multicultural nature of the candidates, policies must be developed which make allowances for the powers of expression used by the candidate using non-standard English. A step in the right direction would be deformalization of the language used in the teaching and assessing of examination courses at 16+. The Bullock Report commented:

> No child should be expected to cast off the language and culture of the home as he crosses the school threshold, nor to live and act as though school and home represent two different cultures ... The curriculum should reflect many of the elements of that part of his life which a child lives outside school.[19]

Any teacher working with pupils from varying ethnic backgrounds would notice that often the child does indeed cast off (or attempts to cast off) the culture – and in particular the language – of the home in exchange for acceptance by peers and/or teachers who adhere to a curriculum language which does little to head off 'culture clashes' or crises of identity. The role of language in the science curriculum needs to be discussed in a wide forum, taking due note of the views of the minority ethnic communities, in the hope that science educators will eventually adopt a more accessible and universally agreed language register.

4 Multicultural education through the medium of Biology

The study of genetics and race

The term 'race' is problematic. It has little genetic basis, but myths about racial differences persist, shaping attitudes and social structure. Social Science lessons can include a rational discussion of race, culture and ethnicity; it is vital that the Biology lessons offer some information about human genetics and an introduction to the complex interaction of inheritance and environment in 'racial' characteristics.

> Although the acceptance in school life of immigrants does not necessarily lead to their acceptance in after school life, what is done in the schools is vitally important.[20]

Children come to school with ideas of both science and pseudoscience and the school is a crucial agent in developing within the child a phenomenology which explains the diversity of human beings. Unfortunately, racism is a set of uninformed judgements, believing, for example, that Negros are less intelligent than Caucasians and that the Chinese are superior to occidentals. Such views are hangovers from nineteenth century pseudoscience. Children find that their attitudes are often challenged by the content and experimental approach of the science lesson; if racist ideas which the child sees as having a scientific basis are *not* tackled, then the school has failed. The National Front waged a campaign outside school gates during the academic year 1977/78. Their newspaper for young people, *Bulldog*, contained no fewer than three articles on race and ethnicity in the first two issues, on 'the race problem', 'race mixing', and 'the Jews'. Their 'facts' about race state that:

> The claim that integration is harmless and that 'all races are equal' is totally false. Races differ, not only in the colour of their skins, but in other physical ways and especially in temperament and innate intelligence.
>
> Scientists tell us that these differences are not the result of environment only, but mainly of heredity. 'Men are not born equal' says Professor Francis Crick, a Nobel Prize winner ... It is not bigotry to oppose multi-racialism. It is a natural healthy instinct to preserve one's own kind. A leading scientist, Sir Arthur Keith, has said that what is called prejudice is simply Nature's way of improving Mankind through racial differentiation.[21]

It is clear from the above passage that weight is lent to the argument by using scientists to justify the 'truth' of the statements. As educators, we have to be aware of the pervasiveness of this type of argument, and we must combat the genetic argument through the Biology curriculum. An alternative view is given in the following quote from an article written by practising teachers for teachers:

> Studies in Biology should provide an appreciation of the genetic basis for differences and similarities between human beings, and should promote an awareness that in biological terms the concept of race is meaningless for human populations.[22]

Australia, a country with a high proportion of recent settlers, is reconsidering the position of its cultural minorities; Claydon *et al.* have commented on the role of science in discussion of ethnic differences:

> The genetic inferiority thesis is not new to Australian science literature and the effects of the debate permeate through all levels of society ... The quality of education can only be improved if debates around such ideas are organised in a structured manner within the classroom ...[23]

In Biology, race is often treated as unproblematic. Brues,[24] writes that the concept of race is a division of a species by the frequency with which certain hereditary traits appear among its members, an example of which might be hair colour or type. Race is also often associated with geographic areas, and this is shown by the images that are presented in many standard Human Biology texts, which identify Africa solely with Negroes. There is little reference to the infinite variety of ways in which division and sub-division can be made, for example, that in Europe there are a number of races and not *a* European race. Very often variation within race is not considered at all. Within the scope of multicultural education, a first question is whether it is feasible and/or desirable to define an individual race at all.

Together with notions of race, there are ideas of racial purity. Some physical anthropologists have sometimes used the terms 'primary' and 'secondary' races in order to explain populations and types. A primary race is defined as a fundamental extreme and a secondary race as the intermediate between neighbouring populations.[25] This model is put forward in the same way as the model which explains the existence of hybrid forms of *Biston betularia*, a peppered moth which exists in a number of morphs. Such ideas of racial purity are linked to questions such as: Have races been purer in the past? The answer given by many physical anthropologists has been that in the past there were fewer people: therefore racial groups were more isolated, and this resulted in greater differentiation. Archaeological research has shown that the history of people is one of migration and movement at various times, but this is rarely cited in Biology texts dealing with human populations; the usual approach is that race is not unique to people and that geographical variation in animal species tends to result in the formation of sub-species, with discernible differences.

The differences ascribed to people are biochemical, pigmentary, morphological, etc., and the inherent danger in this is that the idea of race may lead people to seek differences to prove hypotheses. For some, the purpose of such investigation is to establish degrees of superiority/inferiority, which explain the status quo, or to provide reasons for under-achievement.[26] There is frequent reference to studies by

Ferguson,[27] who in 1914, 1917, and 1919, obtained scores for cognition tests of Negroes and various hybrid Negroes. He showed that quadroons scored 94 per cent of the mean mark of their Europid counterparts, whereas full Negroes scored only 73 per cent of the mean marks of the Europids. This would appear to show inheritable differences in cognitive ability between the two groups independent of any environmental factors. However, according to Rogers,[28] there is no reference to the existence of a colour hierarchy, which would affect social status or class of individuals and hence their linguistic skills and ability to read.

Ferguson's studies in the first and second decades of this century mark the starting point for a body of research which has become part of educational theory, and indeed, part of the pseudoscience of racism. This research includes the work of Jensen[29] with his arguments about social/genetic inferiority, later supported by Eysenck. Here Biology and the associated syllabuses and examinations have an important role to play in the anti-racist education of pupils, by taking them through some of the evidence (see, for example, Gottesman[30]) that shows the effect of environment in regulating the expression of genes. This is usually crudely posed as the 'nature versus nurture' argument in explaining ethnic differences in raw IQ scores. The assumption that intelligence is essentially inheritable and due to genetic expression can be challenged by the use of valuable and interesting examples such as those cited in Whitten and Kagan.[31] The use of science to justify ideas of racial and cultural superiority should be confronted by full discussion within the Biology curriculum.

Ashley Montagu argues that it would be best if the term 'race' is abandoned. He agrees that factors which cause speciation in other animal groups have also been operative in the human species — but with an added difference, namely, the consequences which have resulted from people's entry into that unique zone of adaptation, the fabricated part of the environment — culture. Montagu adds that:

> The man on the street uses the term in much the same way as it was used by his 19th century compeer. Here physical type, heredity, blood, culture, nation, personality, intelligence and achievement are all stirred together to make the omelette which is the popular conception of 'race'.[32]

Montagu suggests that the term 'genogroup' might be more appropriate. He defines a genogroup as a breeding population which differs from other breeding populations of the species in the frequency of one or more genes. Montagu concludes that the term genogroup gets as near to a statement of the facts as a term can, whereas race goes far beyond the facts and only serves to obscure them.

Multicultural themes in Biology

The question of race, genetics, and intelligence are not the only biological topics which are seen to have multicultural relevance. Health, human population, world resources, disease and diet are other examples. When considering these themes pupils should be encouraged to recognize that each society has its own values, traditions and styles of everyday living, which can be studied in a comparative manner.

These are already major topics within the traditional Biology syllabus. The main problem is the scarcity of material available to teachers to supplement and service these curriculum areas with a multicultural perspective. The experience of individual pupils is an untapped resource, and an input from a culture 'represented' by a pupil, may also help increase that pupil's motivation and participation. This is also seen as 'good practice' by both the ASE[33] and the DES.[34] The ASE states that what it considers to be a good course is one which would 'allow youngsters to gain a sense of social meaning and identity as well as personal autonomy'. This embraces the sentiments of Jeffcoate,[35] and implies that multicultural education not only benefits pupils collectively, but also has an important contribution to make in the personal development of the child from the minority ethnic group. The themes outlined above should be considered not only within the secondary curriculum, but also in the science curriculum of the primary school as part of the work that children do in considering their self-image. Thus, if increasing the horizons of the Biology curriculum increases motivation, interest, and participation, then multicultural Biology is likely to be good Biology. This view is supported by the DES in their survey of secondary schools:

> Good schemes of work were widely based with depth of material and guidance ... A question of some interest is the degree to which school science courses include up to date and relevant practical applications of the concepts and principles being studied. Applications can concern the everyday life of the pupils ...[36]

The DES were concerned with the stimulus material being used by science teachers; and they reported that immediacy was linked both to the familiarity of the examples used to illustrate concepts, and to the construction of novel learning situations. Multicultural themes in Biology represent social aspects of science which are concerned with the everyday life of pupils. The use of such themes offers Biology teachers an opportunity to develop new resources, and could also bring the importance of multicultural education through Biology to the attention of the examining boards.

II DEVELOPMENT OF BIOLOGY SYLLABUSES APPROPRIATE TO A MULTICULTURAL SOCIETY

1 Cultural aspects of Biology

The previous chapter discussed the need for science syllabuses and examinations which are culturally fairer, and which reflect cultural diversity. It also suggested themes in Biology which are particularly relevant to multicultural education. Biology offers considerable scope for including topics which can educate children about the origins and natures of other cultures. This section outlines such broad topic areas, and considers how they might be extended to have a culturally wider perspective. It draws on the views of the DES,[37] the Community Relations Commission[38] and the Haringey Multicultural Science Group of which the author is a member. (See also Appendix C which presents this information in diagrammatic form.)

Suggested topic outlines

a. Plants and animals

Fauna: to include animals of significance from countries represented by pupils from minority ethnic backgrounds, but not to replace those examples already used. Economic and social issues covering the distribution and possible uses/importance of animals to be included. Ecological balance of other populations.

Flora: similar to above, but here more emphasis should be placed on the economic importance of certain groups or species, e.g. millet, maize.

b. Food and nutrition

Nutrition: an examination of plant storage organs familiar to groups outside that of the majority ethnic groups, e.g., cassava, yam; consideration of the context of food chains (to include a portrayal of human beings as omnivores); increase the variety of examples used in examining animal nutrition especially those aspects of contemporary human dietary practice.

World food problem: malnutrition to be viewed not only from a Western point of view, but also take into account experiences from the Third World: starvation, obesity, problems of diet in the Western developed societies; the effectiveness (or otherwise) of Western aid; the relationship between energy requirements, nutritional requirements and dietary practice, e.g., vegetarianism.

Oral hygiene: tooth decay/caries, fluoridation, methods of cleaning teeth.

Parasitism: examples to include a comparison of important parasites in different environments, e.g., the flea, tapeworm, significance of nematodes, other crop parasites and their effects on economies.

c. Respiration: respiratory adaptations to altitude and temperature (including the role of humidity).

d. Transport: adaptation of circulatory system to temperature of environment, origin and significance of blood groups, sickle-cell anaemia and its origin, thalassaemia.

e. Growth: a comparative view of growth rates, the effects of diet on growth; consideration of the differences in 'race' and the onset of puberty, interacting with dietary factors.

f. Homeostasis: the incidence of diabetes in various ethnic groups and its relation to diet; how water and salt contents change in different diets.

g. Reproduction: a world perspective on childbirth, including methods of delivering babies, child care, and birth control.

h. Genetics and variation

Genetics: a sensitive consideration of genetic-related disorders, e.g., sickle-cell anaemia, and their incidence within racial groups; a careful and discursive treatment of IQ testing and the related debate about the influence of genes and environment on the development of the individual.

Variation: a consideration of the possible origins of 'races' as hypotheses; biological definitions of 'race' and their validity; human morphology and its significance in relation to the environment; variation within racial groups.

i. Environmental issues

Conservation and pollution: resources and their differing uses; economic, social and political implications; the relation of culture to resource use; conservation issues of a global dimension.

Disease and health: diseases of global significance, e.g., influenza, yellow fever, herpes; epidemics, pandemics; a comparison of health care in relation to needs and health 'technology'; the notion that what constitutes a healthy body varies enormously between societies; the use of pharmaceuticals in developed and underdeveloped societies; folk medicine and its role, e.g., homeopathy, 'bush teas', acupuncture, etc.

Public health, i.e., health care for large numbers of people, differences in structure and provision; the World Health Organization.

To include these topics in an already crowded but not comprehensive Biology syllabus for the fourth and fifth years of secondary education may seem to be irrelevant to the discipline as it stands. It might be argued that a culturally-neutral syllabus with corresponding examples could be developed as an alternative. However, it is being suggested here that 'good' Biology is multicultural Biology and that it serves two purposes within what pupils view as being part of the most important two years of schooling. First: pupils from minority ethnic communities need to be educated to take up their lives with an awareness of their own culture; without this education, their view of themselves and their history may soon be expressed in feelings of alienation and low personal esteem. Second: pupils from the majority culture may have their views and prejudices of other cultures confirmed or modified by the content of the curriculum. The examination curriculum has a high status, and a narrow curriculum, which does not take into account the cultural diversity within schools and society at large, may not do enough to challenge racial and cultural prejudice.

Cultural fairness can only be established if an exchange of cultural information is given a high priority and status within the curriculum and especially within the examination curriculum. A closer look at the topics outlined above indicates that their inclusion would not take up too much time, and that they would not be competing with other areas of the examination syllabus but would be an extension of them. They are part of a whole view of Biology which includes variation as well as interrelations within the environment.

2 The draft national criteria for 16+ examinations in Biology

Multiculturalism as a subject-independent examination feature

The Schools Council Working Party on national criteria for the proposed single system of examining at 16+ identified several features of examinations which are independent of subject. One suggested criterion is that:

> Examinations should be designed to minimise risks of bias with regard to sex and ethnicity and to reflect positively the multicultural nature of society.[39]

The Schools Council's Science Committee's 16+ Working Party then recommended that:

> Each syllabus should draw widely on situations and examples from different strands within contemporary society, in order that it should be accessible to pupils of both sexes and of different cultural backgrounds. Assessment procedures would not be valid unless they were based on the same principles.[40]

Current proposals

The GCE and CSE Boards' Joint Council for 16+ National Criteria made public their draft criteria for 16+ examinations in Biology,[41] and these are reviewed here to see whether they are appropriate for assessment in a multicultural society.

Aims. The series of aims outlined, although full, do not provide clear guidance for the teachers who wish to know how they might adapt the aims to suit their own teaching situation. Included as an aim is the development of an appreciation of the significance of Biology in a social context. This aim does not represent a radical departure from similar views expressed by boards in their outline of aims of current examinations, but it is a justification for including several topic items which were listed earlier in this chapter.

Assessment objectives. The assessment objectives proposed are not ethnocentric unless the examples, biological facts, or information are culturally loaded. This will depend on the content and structure of the actual examinations, including the choice of questions. However, it would be useful to see how the boards interpret the following assessment objective:

> To demonstrate knowledge and understanding of ... personal, social, economic and technological applications of biology in modern society.

It is the view of the author that 'modern society' might be understood to mean 'this' society, by those teachers who may not recognize the need for all children to be educated for life in a culturally diverse society. It might be better to reword the assessment objective as '... applications of Biology in this and other societies'. Given this wider scope, a larger number of syllabuses might well be developed, each providing education on principles but differing in the subject matter.

Content. The Joint Council proposed a minimum core content, the intention being to allow interpretation and extension as desired by the regional 16+ examining groups. However, there is a recommendation that:

> the content is interpreted through practical studies of a broad range of organisms and related, wherever possible to human and social biology and to developments in biotechnology.

The addition of a further clause, such as 'and to the needs of a multicultural society' would clarify the issue for both boards and schools.

Relationship between assessment objectives and content. The proposals suggest that:

> At least 15% of the marks in any complete examination must be allocated to topics related to the personal, social, economic and technological applications of biology in modern society.

It now behoves those working for the examining boards to outline how they will meet this suggestion. It would also be helpful if the language implications raised by the Assessment Objective which refers to the use of 'appropriate terminology' are made clear by outlining the meaning of this statement. It is important to candidates from cultural and linguistic minorities that examiners continue to accept a wide variety of language structures. As most candidates for 16+ examinations will be non-specialists, the informal language (as is used in television documentaries or scientific literature for non-specialist readers) may be more appropriate, and would more readily accommodate the wider range of language of ethnic minority pupils. Different considerations might of course apply at A-level and beyond, in the assessment of science specialists.

Techniques of assessment. The techniques of assessment proposed involve three components: school or college-based assessment, written papers and practical tests. The future of the essay question demanding extended prose is unspecified. The DES commented that:

> The staff of one science department had noted over several years that language difficulties had severely handicapped examination candidates in science subjects and particularly ethnic minority group pupils. Despite their proven ability in practical science and their wide background knowledge these pupils had difficulty in presenting their knowledge in the form of essay answers required in formal examinations.[42]

This point was also stressed by the Rampton Report. The move away from extended essay questions is welcome. The effects of language difficulties can also be minimized by basing more questions on a visual presentation where diagrams and graphic/pictorial representation of data are used. The use of oral questions is not recommended by the Joint Council as doubt is raised about their reliability. Practical examinations and assessed practical coursework are perhaps more relevant in science than are oral examinations.

General. No guidelines are given as to how the sentiments expressed in the sections concerned with the personal and social aspects of Biology are to be incorporated into a culturally-fairer scheme for setting and marking examination questions. Alternative criteria for acceptable answers should be indicated in examiners' reports, and discussed at teachers' meetings.

3 Existing 16+ Mode I syllabuses in Biology

Earlier in this chapter syllabus topics were listed which would allow a multicultural approach. A review of all the O-level and CSE Mode I Biology syllabuses over the years 1980-83 showed that they all covered very similar areas. The syllabuses were analysed for their coverage of topics which allow a multicultural approach. Table 1 shows that most syllabuses offer most of these topics; however, the topics which feature least frequently include those which could relate most directly to multicultural Biology, e.g., evolution of Homo sapiens, race and distribution; indeed race does not appear overtly in any of these syllabuses.

Of the twenty* boards reviewed, five see it as important to include as an aim for the examination, the development of an awareness of the social importance of Biology, and only two boards see this as an ability to be tested. AEB statement on 'Aims and Objectives' says that candidates should learn 'to appreciate the value and use of biological facts to human existence', and that this should be examined. Although some examination boards may regard it as important for the pupils to be aware of the social importance of Biology, they do not usually address themselves to topics of social significance. Sixteen such topics are included but not examined, the teacher may well gloss over or omit them, and so ignore this multicultural aspect of Biology.

'The variety of living organisms' is covered by all the examination syllabuses reviewed, but the selection of examples is sometimes restricted. LR notes that wherever possible the organism(s) should be studied in the context of habitat, and this does not preclude an important plant such as sugar-cane from being examined as a monocotyledon; however, other boards place constraints on the examples used, prescribing those which are most commonly accessible. Perhaps schools could be left to determine which examples are used to illustrate the variety of living organisms, so that the composition of a school population together with any special interests and local factors could influence the choice of materials.

---

* WJEC is treated as two boards: CSE and GCE. For abbreviations of boards, see Appendix A.

Table 1 Current Mode I O-level and CSE syllabuses and topics which allow a multicultural approach (1980-83)

x = included in syllabus; - = not included

Board*	Variety	Storage of food materials	Food chains/webs	Foods and diet	World food resources	Teeth and tooth care	Parasitism	Respiratory adaptations	Blood and blood groups	Blood sugar levels	Water and salt regulation	The skin	Birth control	Birth and post natal care	Genetics	Human variation	Genetic diseases	Wo/man and environment	Disease and public health	Evolution of Homo sapiens	Food production	Food preservation	Race and distribution
AEB	x	x	x	x	x	x	x	-	x	x	x	x	x	x	x	x	x	x	x	-	-	-	-
ALSEB	x	x	x	x	x	x	x	-	x	x	x	x	x	x	x	x	x	x	x	-	-	-	-
Cambridge	x	x	x	x	-	x	x	-	x	x	x	x	x	-	x	-	x	x	x	-	-	-	-
EA	x	x	x	x	-	x	x	-	-	x	x	x	x	x	x	x	x	x	x	-	-	-	-
EM	x	x	x	x	x	x	x	-	x	-	x	x	x	x	x	x	-	x	x	-	-	-	-
JMB	x	x	x	x	x	x	x	-	x	x	x	x	x	x	x	x	x	x	x	-	-	-	-
London (040)	x	x	x	x	-	x	x	-	x	x	x	x	x	x	x	x	x	x	-	-	-	-	-
LR	x	-	x	x	x	x	x	-	-	-	x	x	-	x	x	x	x	x	x	-	x	-	-
NR	x	x	x	x	-	x	x	-	x	x	-	x	-	x	x	-	-	x	-	x	-	-	-
NW	x	-	x	x	x	x	x	-	-	x	x	x	-	-	x	x	-	x	-	-	-	-	-
O and C	x	x	x	x	-	x	x	-	x	x	x	x	-	x	x	x	x	x	x	-	-	-	-
Oxford	x	x	x	x	-	x	x	-	x	x	x	x	-	x	x	x	x	x	x	-	-	-	-
SE	x	x	x	x	-	x	x	-	-	x	x	x	-	x	x	x	-	x	x	-	-	x	-
SR	x	x	x	x	x	x	x	-	x	-	x	x	x	x	x	-	-	x	x	-	-	x	-
SUJB	x	-	x	x	x	x	x	-	x	x	x	x	-	x	x	x	x	x	-	-	-	-	-
SW	x	x	x	x	x	x	x	-	x	-	x	x	x	x	x	x	-	x	-	-	-	-	-
WJEC (GCE)	x	x	x	x	-	x	x	-	x	x	x	x	-	x	x	x	x	x	-	-	-	-	-
(CSE)	x	x	x	x	x	x	x	-	x	x	x	x	x	x	x	x	-	x	x	-	-	-	-
WM	x	-	x	x	x	x	x	-	-	x	x	x	-	x	x	-	-	x	-	-	-	-	-
YR	x	x	x	x	x	x	x	-	x	-	x	x	x	x	x	x	x	x	-	-	-	-	-

All the boards provide scope for testing foods for the presence of nutrient class, e.g., starch. No Mode I syllabus fixes or recommends the type of food to be tested or used as an example. As one teacher commented:

> It never occurred to me to start testing anything other than a potato for starch until this boy brought out rice from his lunch box ... It was an immediate success.

Eighteen boards mention 'balanced diet' as part of their animal nutrition topic, seemingly accepting the problems of variation in diet between cultures. Six of the boards include diet deficiency diseases as part of their syllabuses and these represent the most 'tangible' multicultural element within the present examination courses. LR attaches this to the general problem of world food resources. The majority of boards do not have syllabuses which allow the question, 'Why do dietary deficiency diseases arise?' to be answered within the broader view of global resources.

SR includes the care of teeth as part of their nutrition topic without specifying fluoridation, brushing, etc. This represents a positive step, as oral hygiene has a large number of culture-dependent practices and these can be related to the incidence of dental caries.

Homeostasis features in all of the O-level and CSE syllabuses analysed. Emphasis is placed by the CSE boards on the effects of hormones (excess and deficiency) and the function of the kidney. Seven of the GCE boards allow the use of diabetes to illustrate malfunction of the mechanism of homeostatic regulation. The prevalence of diabetes among some of the minority ethnic groups with high carbohydrate intakes, gives this topic social and cultural relevance and is easily backed up by materials from some community health councils.[43]

The function of the skin as an organ of temperature control in the mammal is a part of all O-level and CSE syllabuses; however, melanin is not included in any of the syllabuses as an explicit point, even though its role in the prevalence of rickets in black and brown children in Britain is well known. Pigmentation also represents an example of continuous variation within Homo sapiens. Only two syllabuses give examples of variation of human form (both continuous and discontinuous variation), and the examples used (height, tongue rolling) cross sex and racial groups.

'Transport systems' in animals (blood and circulation) is usually in the syllabus, for example:

> An elementary study of the mammalian heart: outline of the arterial and venous systems: circulation of blood and lymph and their functions as agents of transport. (AEB)

Other syllabuses such as LR's and WM's include aspects of health in the section on blood and circulation. Sickle-cell anaemia and thalassaemia are blood disorders which are particular to cultures and races, but only heart disease, a circulatory disorder, is included within the syllabuses, even though many people of West Indian, African, Asian and Mediterranean origin in Britain are hererozygotes and carry the 'sickle-cell trait'.

The life history of Homo sapiens figures in nineteen of the syllabuses with reference to birth and parental care. This aspect of the syllabus readily lends itself to a comparison of child rearing practices across cultures showing that there are greater similarities in the manner of post-natal care than pupils at first realize. These topics can be neatly integrated with those syllabuses which include public health, immunity and vaccination. Birth control and its implications for the world's population, are seen as examinable by only eleven of the boards. AEB sees it as important for the candidate to understand 'the necessity for control of the size of the human population related to world food supplies ... [which should be] seen as one of the greatest human problems of our time'. Although this is a welcome inclusion, it is left to the teacher to decide on the sensitivity of the topic, the manner in which it should be approached, and level to which it is developed. Questions about the distribution of food resources in the world could be particularly relevant here.

The examining boards differ in the importance they attach to inheritance within their syllabuses, and the examples used to illustrate heredity. Some boards specify the practical examination of *Tribolium* and *Drosophila* to illustrate the principles, whereas others use inheritance of eye colour in humans as an example. Extension of the examples that could be used by the teacher would serve to make these topics more relevant to the multi-ethnic community. In its discussions about genetics and inheritance, the Haringey Multicultural Science Group suggested sickle-cell anaemia, skin colour, and hair type as being characteristics which should be part of any course in Biology for fourth and fifth-year pupils.

YR is the only board to include the evolution of man in its syllabus, and it leaves the detail and approach to the teacher; the rest of the boards have not included this within their syllabuses, despite the fact that children always ask about the origins of human variation. It may be that the boards do not wish to get drawn into the argument that has raged in the USA, between biologists and the creationist lobby. However, the examination curriculum has a duty to allow that argument to take place in the classroom as it is so often bound up with 'racist' ideas of the differing origins of groups of humans.

All syllabuses contain a topic which examines the effects of humans on the biosphere, which provides a window through which the pupil can see beyond Britain. London supplements this with a section on methods of conservation. This links itself to an examination of the relationships of people to resources and the environment, and allows the pupil to assess how this changes according to the level of technology and cultural values within a given society. Twelve of the boards include this topic, and they also include disease and public health in their syllabuses, where the significance of some diseases only becomes apparent when a view of a changing (developing) world is adopted. EA includes food preservation in a consideration of the role of micro-organisms in transmitting disease. This part of the syllabus focuses on canning, freezing, salting, dehydration, soaking, and irradiation as examples of preserving methods. The board referred the author to some schools which cover this topic, and one school commented that: 'The few West Indians we have in our school seem to have a higher proportion of salted and brine-preserved foods.' This school goes on to make the point that the inclusion of a topic in an examination syllabus does not necessarily mean that it will be taught, especially when its importance is reduced, in the eyes of the teacher, by being excluded from the examination paper. The choice of topic, and thus the degree of enthusiasm with which it is taught, depends on the examining panel (or chief examiner) and any feedback from the schools. Only a minority of the boards have a commitment to Social Biology, and therefore have topics within their Biology syllabuses which directly foster a multicultural approach. This may be because many boards also offer Human Biology, and Human and Social Biology as separate examination subjects, and they see the multicultural aspects as being more properly placed here.

A less detailed review of 16+ Mode I Human Biology and Human and Social Biology syllabuses was also carried out. Three GCE and seven CSE boards offered at least one of these as examination subjects. The incidence of topics which would allow a multicultural approach was very similar to the Mode I Biology syllabuses, but there was evidence of a wider, world view of Biology. Five boards mention world living standards: NW has a section entitled 'Standards of Living at Home and Abroad', fostering a comparative view of the world which is part of multicultural Biology; YR includes race and major human racial groups as part of the Human Biology syllabus, and sickle-cell anaemia is included in the topic on disease.

AEB declares as one of the aims of its Human Biology syllabus (1983) at O level:

> This exam syllabus has been designed as a framework around which schools and colleges can build a teaching syllabus which is relevant to the needs and interests of their students.

A world perspective seems a major omission from this aim: this perspective allows students to place themselves within the context of a number of different environments and to understand different ways of responding to environments. Such a perspective is one which extends beyond a knowledge of content to influence the attitude of student biologists: it is relevant to the needs of all students in this society, and it should be stated clearly within the aims for 16+ examinations. This would augment the world content which is already present to some extent in Human Biology syllabuses.

The points raised here are also being taken up by overseas colleagues. Examinations for overseas candidates are offered by Oxford, Cambridge, and London boards. Cambridge and London both offer special papers for candidates in the Caribbean area taking Human and Social Biology. In 1976, the Trinidad and Tobago Working Party on Education reported to the Prime Minister, that

> There should be a reassessment of the entire content of subjects at the Junior Secondary level with a view to replacing the present GCE. The new examination should allow for a content more applicable to the child, preparing him for his role in a society that is no longer a colony.[44]

The trend in education, at that time, was towards an examination curriculum more suited to the needs of the post-colonial West Indian society. The author wrote to Grenada and Trinidad to assess the present situation; three replies were received concerning Biology examinations. A Trinidad Junior Secondary School wrote:

> We still offer the children the opportunity to take the GCE but we are at the moment trying to get more of them to take our own Caribbean Examinations Council exam in Biology. I feel it is a better exam because it has been developed for use in the West Indies and has a better syllabus, with more agriculture and emphasis on health and hygiene ... very often the old syllabus did not really answer the questions of the children ...

The copy of the Caribbean Examinations Council (CEC) syllabus sent with the letter is dated 1979 and so may well be out of date. It does however include a section on human variation, also covering sickle-cell anaemia. Diet-deficiency diseases are listed, including kwashiorkor, beri-beri, anaemia, diabetes, high blood pressure and rickets. A Grenadian headmaster wrote that the CEC examination was more relevant for Grenadians than the external GCE O-level examinations, but that the latter still carried higher status.

4 Multicultural Biology and 16+ Mode III examinations

Over 100 letters were sent out to schools in sixteen education authorities that are known to have significant representations of minority ethnic communities. Some of these were followed up by telephone calls and further letters. Replies were received from 39 schools, eleven of whom expressed interest in the project but were unable to help or were unable to cooperate with the study. All of the returns were from the large urban cities.

Most of these schools were discontented with the 16+ Mode I syllabus offered by the boards, in terms of content, aims, methods of assessment. One school wrote that since running its Mode III examination, grades had improved significantly and pupils enjoyed the course much more. None of the respondents referred to the multicultural nature of their school as a major reason for adopting a Mode III examination. The following response, from a Brent school, was fairly typical:

> The primary reason for developing a Mode III examination was to create a course that related much more to the needs of our pupils and so improve their chances in the exam. Before this, we had little control over the outcome of exams but we now prepare our pupils much more for the examination because we understand the demands of the course. We were teaching a practical based course and entering pupils for an exam that did not really test their practical ability ... Our school has a large number of Cypriot pupils who had difficulties in writing the extended answer question. Our Mode III exam has multiple choice and short answer questions.

Another school reported that they had discussed multicultural Biology but resources were difficult to find. A community supplementary school that offers extra classes in Biology commented:

> The Mode III exam represents a second-class exam that schools have developed to enter our children for. The sort of content that you are endorsing should be part of mainstream syllabuses in all sciences.

A teacher of this school commented:

> I teach Biology in a school that has a Mode III exam with a traditional content.
>
> So far no attempt has been made to use the experience of the children themselves [referring to the pupils from the ethnic minorities] or to educate them about Biology in relation to their lives.

Other replies concerning the general issue of multicultural Biology are exemplified by this reply from an ILEA school:

> We should be educating children about their similarities rather than their differences.

However, there are a large number of topics and ideas within the examination curriculum which, if ignored, result in education for the cultural majority alone. An ILEA school which has discussed these views argues that education about diversity should be one facet of a multicultural policy and:

> The content of our course has an emphasis towards Human Biology and in class we regularly have discussions about examples that relate to the different peoples of the world.

In addition,

> The problem has occurred in the images that the children get in the standard Biology texts of starving children in Africa, and that all of these places are riddled with diseases. We felt it our duty to challenge these ideas so we introduced some new examples into our academic and practical content.

This school had held a departmental meeting to discuss images in Biology texts and had decided to write some of their own material. The school mentioned that they had been able to do this because the department has a West Indian and an Asian member of staff and could draw on their views and knowledge.

The author's letter to schools also asked them to suggest multicultural items that might be included in their Mode III courses. The replies included references to:

- sickle-cell anaemia, frequency and heredity
- blood groups and race
- heart disease and diet
- world diets, energy values of, significance of, comparison with UK diet
- food tests to incorporate foods of cultural minorities
- diet-deficiency diseases treated alongside obesity
- famine and starvation, world food resources and inequalities
- diseases of the developed and underdeveloped world
- racial variation and adaptation, evolution of races
- pigmentation, its significance
- comparative medicine and hygiene

These suggested items match closely those cited earlier, and are all already present within one or other of the present 16+ Mode I syllabuses. The schools also pointed out a number of topics that they had included as 'good practice'. One school mentioned:

> Crops and world crop production. A comparison of rice, wheat, maize as carbohydrate staples ... The mosquito as an example of a parasite of significance and the manner in which its

control with DDT has been subject to the economic position of countries in the Third World, i.e. financial versus social considerations.

The multicultural topics actually covered in the Mode III syllabuses received were as follows:

energy requirements and actual energy intake in different countries

testing a wide variety of foods (including yam, sweet potato, pasta, green peppers) for starch, fat, sugar, protein, and vitamin C content

oral hygiene and the incidence of dental caries in relation to diet

hay fever and its incidence amongst the ethnic minorities

origin of blood groups

the role of melanin in the skin, carotene

diseases of world-wide significance

how we taste spicy foods

the genetics of sickle-cell anaemia

the distribution of races

racial morphology

From the submissions, it became clear that although most schools had adopted Mode IIIs in order to have a greater degree of control over the syllabus and its assessment, some schools have also begun to introduce components and themes to their syllabuses which will produce examinations which are far less Western in outlook. The purpose of this is summarized by an ILEA school:

> Within Biology exists an opportunity to educate children about the basis for difference and to see that many of the ideas they have about other peoples are stereotypes.

5 The role of project work

Several schools have used 16+ Mode I and III courses that have an extended piece of work as part of the assessment. Though some examination boards have been very critical of project work components, there has been support for this method of assessment from schools provided pupils do not undertake too many. Project work offers the pupils the opportunity to choose an area in which they are interested, and to show how they can relate the content of the syllabus to themselves. The project could allow individuals to investigate, in more depth, culturally or racially specific topics in a way that would not be examinable in a written paper or practical. One Haringey school stated:

> Syllabuses are constraining in the depth to which investigation can be carried out. Multicultural aspects tend to be peripheral to biology syllabuses and are not examined. The content of most syllabuses is so vast that it is difficult to cover all of it in 140 minutes per week over two years. Project work or individualised work would allow such points as you are suggesting to be included within the assessable parts of the exam.

It is possible that the needs of the ethnic minorities could be accommodated by the inclusion of multicultural Biology as one of the aspects of project or coursework. Combined with the assessment of the social aspects of Biology, project work would allow pupils in all areas of the country (not just the cities) to consider these aspects even though they might not appear in the final examination paper in specific questions.

6 Developments in joint 16+ examination syllabuses in Biology

There are at present three joint O-level/CSE Biology examinations available. They are from the Cambridge/EA consortium, from the JMB/YHR*/NW/ALSEB consortium, and from WJEC.

The JMB/YHR*/NW/ALSEB syllabus has a content which reduces the amount of Human and Social Biology offered in their separate syllabuses. The welcome trend within all these consortia is towards an emphasis on the processes in Biology concentrating on testing observational, comprehension, and deductive skills though unfortunately the topics discussed earlier, which are amenable to a multicultural approach in practical work or in content, have not been included.

There will be from 1985, a new joint O-level/CSE examination by the EA/LR/London consortium which includes in its aims 'To appreciate the significance of biological science in social and economic developments'. This proposed syllabus is specific about the content of Social Biology. The group has produced a detailed syllabus and specimen examination papers, but these specimen papers do not seek to examine any of the social aspects of Biology, and remain a disappointment. The lack of school-based assessment tends to offset the more positive aspects shown by the syllabus, where a section on diet includes an examination of roughage, a balanced diet, daily energy requirements, obesity and dangers of food additives. World food supply is detailed as

> Problems leading to famine; unequal distribution, climatic factors, increasing population.

Dental hygiene is included in terms of proper care of teeth, dietary considerations, and the use of fluorides. A comparative treatment of this topic would be welcome and consistent with the experiences of some candidates as well as being of great benefit to all examinees. There is also a lengthy section on reproduction in humans and comparative treatment of parental care and feeding. Sickle-cell anaemia is included, as an illustration of change in a gene, in the genetics component of the syllabus. These are all useful contributions to multicultural Biology and it is hoped that the consortium will follow up this welcome development with an attempt to find ways of assessing this material.

---

* WYL and YR merged to form YHR (Yorkshire and Humberside Regional Examination Board) only about two years ago. It may be, therefore, that examinations (and, particularly, syllabuses) relate to the separate WYL and YR boards.

III EXAMINATION PAPERS IN BIOLOGY

1 An analysis of examination questions

The questions from 16+ Mode I examination papers for the four years 1979-82 from the AEB, Cambridge, JMB, London, Oxford, SUJB, ALSEB, EA, EM, LR, NW, SE, SW, WM, YR boards were analysed together with the examiners' reports wherever available.

The questions can be grouped into several types. First there is the long and short essay-type question which requires both the presentation of facts and argument, as well as the illustration of points using diagrams. These questions are confined to the O-level examinations. Secondly there are multiple-choice questions which occur in a variety of forms, and where the 'correct' answer is chosen from four or five alternatives. In the third, and by far the largest, category, are the questions that require short answers involving the interpretation of data, or responding to specific questions, or labelling diagrams. Two boards use comprehension questions as part of their examination; this involves reading a piece of information, and answering open and directed questions on the passage.

The questions were scrutinized for cultural neutrality, cultural bias, and for multicultural aspects of Biology. A culturally-neutral question is one where the required answer is available to all candidates independent of their cultural experience. A culturally-biased question would be one which did not allow equal access to the answer for all candidates irrespective of their cultural background. Questions considered to be testing a multicultural aspect of Biology were those which examined or sought to examine an aspect of Biology related to the topics outlined in Chapter II. In addition, some were picked out as examples of questions which could be expanded to embrace multicultural Biology. Out of the 297 essay-type questions examined, 232 fell in the 'culturally-neutral' category. A typical 'culturally-neutral' question is reproduced below:

(a) Give a balanced equation which summarises the process of tissue respiration

(b) What are the differences between aerobic and anaerobic respiration?

(c) Describe an experiment which demonstrates that germinating seeds give off heat.

(d) How does a named protozoan obtain its oxygen supply?
(London, Jan. 1982)

It should be noted that the above questions allow candidates the freedom to choose their own example.

However, the phraseology of many of the essay-type questions could be difficult for pupils whose English is their second language. For example,

> Explain how muscles contribute to movement of the limbs and gut and to accommodation and iris movements within the eye of the mammal.

The question might be accessible to all candidates if it were rephrased thus:

> Muscles help to bring about a movement of the limbs and gut. In the same way they contribute to accommodation and iris movements within the eye of the mammal. Explain how they do this.

Over 260 multiple-choice questions were examined, and the point already made about language applies to a great many of these questions too, especially where the candidate has to establish the correctness of two statements, and then decide if the second is a correct explanation of the first. A Greek-Cypriot pupil in the fourth year of a Tottenham school picked out two examples which he found very difficult. The examples are hard anyway, and the language may have further mystified this pupil who was born in Britain but still speaks Greek-Cypriot at home:

A synapse is a feature of all reflex arcs in mammals.	Even the simplest reflex arcs involve at least two different neurones.
The volume of gas in the swim bladder of a fish increases with depth.	The density of a bony fish increases as it swims deeper.

These questions ask the candidate to understand the coding key, assess the statements and then decide about meaning. What is required here is not just a knowledge of terms or biological facts, but also a familiarity with complex grammar.

Some multiple-choice questions are culturally neutral, for example:

> Which of the following would yield the largest amount of energy per gram when respired? A, Fat; B, Starch; C, Cellulose; D, Protein; E, Sucrose (London, 1982)

A number of multiple-choice questions contained unnecessary cultural bias, as they asked for answers dependent upon knowledge of examples that were more readily available to candidates from the majority culture (these examples had not been specified in the syllabus). For example:

> Which of the statements below is NOT true of the following food chain? MICROSCOPIC ALGAE - SHRIMPS - HERRING - COD - MAN
>
> A. The microscopic algae are an essential part of the chain
>
> B. The plants are more numerous than the animals
>
> C. All the animals are predators

D. The energy eventually obtained by man comes from the sun

E. There must be fewer cod than herring

The analysis of the short-answer questions common to both the GCE and CSE boards yielded some of the most encouraging signs. For example, this question appears as part of a longer question concerning the female reproductive system,

> Babies in Western countries are usually weaned (introduced to solid food) at an early age.
>
> What specific dietary materials should be provided by this transference from milk to a more varied diet?
>
> (SR)

The board has made the statement very clear. If the word 'Western' had not been used, this question would have been culturally-biased against the experience of some pupils.

Conversely, some short-answer questions use examples that may be unfamiliar to some pupils:

> Some fruits such as strawberries, cherries and raspberries, are large and brightly coloured ...
>
> A sandwich containing wholemeal bread, meat and fresh salad is likely to be more nutritious than a sandwich made of white bread and meat paste ...

The problem occurs when the syllabus does not spell out the examples that will be used in the examination. Imagine the uproar if chapatti or pitta was used instead of wholemeal or white bread and not specified as part of the syllabus. A similar issue is shown in these questions from a multiple-choice paper, which uses culture-specific slogans:

> Relate each of the following slogans to one of the dietary requirements shown below:
>
> 'Carrots help you see in the dark'
>
> 'Brown bread is good for the nerves'
>
> 'Cod liver oil builds bonny babies'
>
> Vitamin A, Vitamin B, Vitamin C, Vitamin D, Iodine

The question might have been easier if the language were more neutral, i.e., less colloquial:

> 'Carrots help to prevent night blindness';
>
> 'Brown bread is essential for a healthy nervous system'; and
>
> 'Cod liver oil builds babies' bones'.

Questions which use proper nouns, need not always use English names, as shown in this question below. Britain has names that originate from many countries and these should be used where possible. For example, Hanife could be substituted for Jeanne and Stefan for Ron in the question below.

Questions 23 and 24 refer to the following diagram.

Below is a family tree showing the inheritance of the ability to taste a certain substance. The gene for the ability to taste this substance is dominant to that for the inability to taste it.

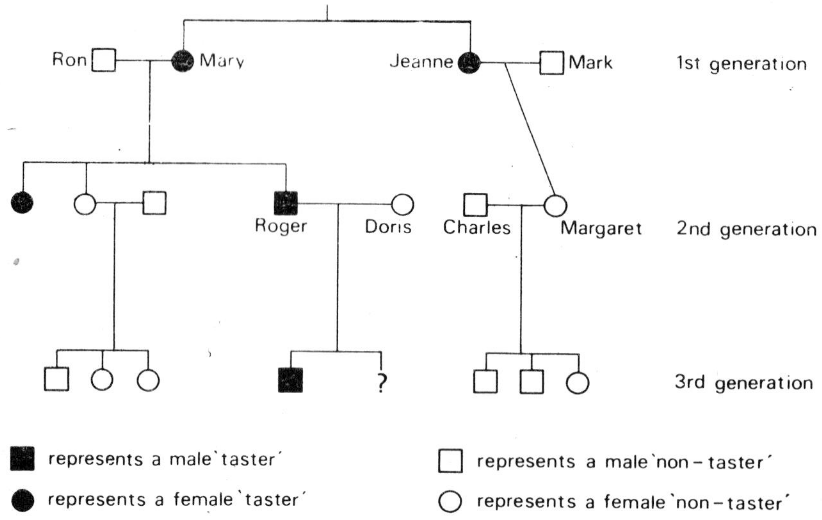

23 What can be deduced about the genotypes of the sisters Mary and Jeanne?

A Mary is homozygous and Jeanne is heterozygous.

B Jeanne is homozygous and Mary is heterozygous.

C They are both heterozygous.

D They are both homozygous.

2 Opportunities for multicultural responses

The following question on the skin has scope for a multicultural answer if the pupil has access to that particular knowledge. The diagram below shows a vertical section of the skin.

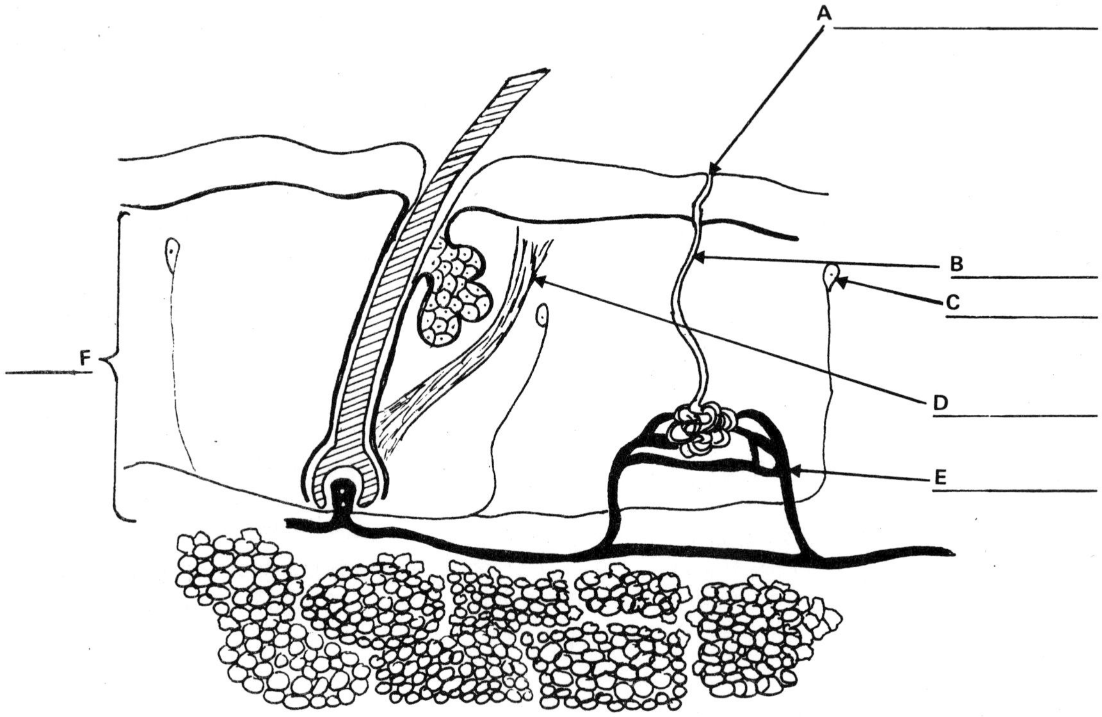

(a) (i) On the diagram label the parts A to F.

(ii) Using the letter X, indicate which layer could be regarded as a food store.

(iii) State THREE functions of the skin (apart from food storage).
(ALSEB, 1982)

Protection from mechanical damage, acting as a barrier to pathogens, retaining water, acting as a surface for temperature control through sweating, the erection of hairs, synthesis of vitamin D, protection from ultraviolet radiation and being a surface for emitting radiated heat (i.e. black body) and therefore cooling the individual's body temperature; these are all examples of possible answers to the three functions asked for by the question.

The following questions, similarly, offer candidates the opportunity to bring in knowledge of multicultural Biology:

With the aid of sunlight the skin can make

A. sebum
B. adrenalin
C. sugar
D. urea
E. vitamin D

(SR, 1981)

Man often tries to develop variations shown by plants and animals because they are useful or pleasant

(i) Give ONE example of variation which has been developed by Man to his own advantage

(ii) Explain how the development of this particular variant may have been brought about.

(WM, 1982)

Give TWO different examples of evolutionary change brought about by Man's activities which are thought to be disadvantageous to Man. (WM, 1982)

The last two questions almost invite the pupil to write from their own experience of Biology. The problem then rests with the examiners and their familiarity with World Biology. The WM Biology syllabus has a larger Human and Social Biology input than many of the other syllabuses.

The following question represents a model and nearly unique example of how multicultural aspects of Biology might be examined:

Normal red blood cells are shaped like a disc. Some people possess red blood cells which are sickle-shaped. These are known as sickle cells. Malaria frequently occurs in those parts of the world where people with sickle cells live. Malaria can be a fatal disease.

A sample group of people, some of whom had sickle cells and others did not, were tested to see whether there was any relation between the presence of malarial parasites and sickle cells.

Table I

	Malarial parasites present %	No malarial parasites %
Sickle cells	28	72
No sickle cells	46	54

(i) What does the data in Table I suggest about the relationship between presence of malarial parasites and sickle cells?

(ii) Give an alternative explanation for the results.

Two volunteer groups were taken, one possessing sickle cells and the other normal blood cells. Each individual was artificially infected with malaria and their health checked for the next 40 days.

Table II

	Developed malaria	No malaria
Sickle cells	2 very mild attack	13
No sickle cells	14	1

(iii) What does the data in Table II suggest about the effect of the presence of sickle cells?

(iv) What would you expect to happen to the proportion of people possessing sickle cells in a population living where malaria is common?

(v) What could happen to the malarial disease, in that area, if man were the only suitable host?

(WM, 1982)

The question shows how multicultural Biology can be examined as a mainstream item. Firstly, the candidate is invited to search for the relationship between malaria and the incidence of sickle cells. Hypothesis is encouraged, and then the candidate is directed to making some predictions about the relationship between humans and malaria.

A small number of 16+ Human Biology papers from seven boards were assessed for their questions which examined world and cultural aspects of Biology. Out of 74 questions considered, all the seven boards offered some good practice. Eleven questions actually attempted to examine the pupils' understanding of the world issues of Biology or used a non-ethnocentric example. One such question is reproduced below.

The table below shows the average life expectancy at birth of people in different countries for the years 1930 and 1970.

Country	Life expectancy 1930	Life expectancy 1970
UK	58	71
USA	59	70
Netherlands	64	74
Egypt	36	52
USSR	48	70
India	26	46
Chile	36	63

(a) (i) Some of the figures have been plotted on the grid below. Complete the histogram (block graph) by plotting the remaining figures.

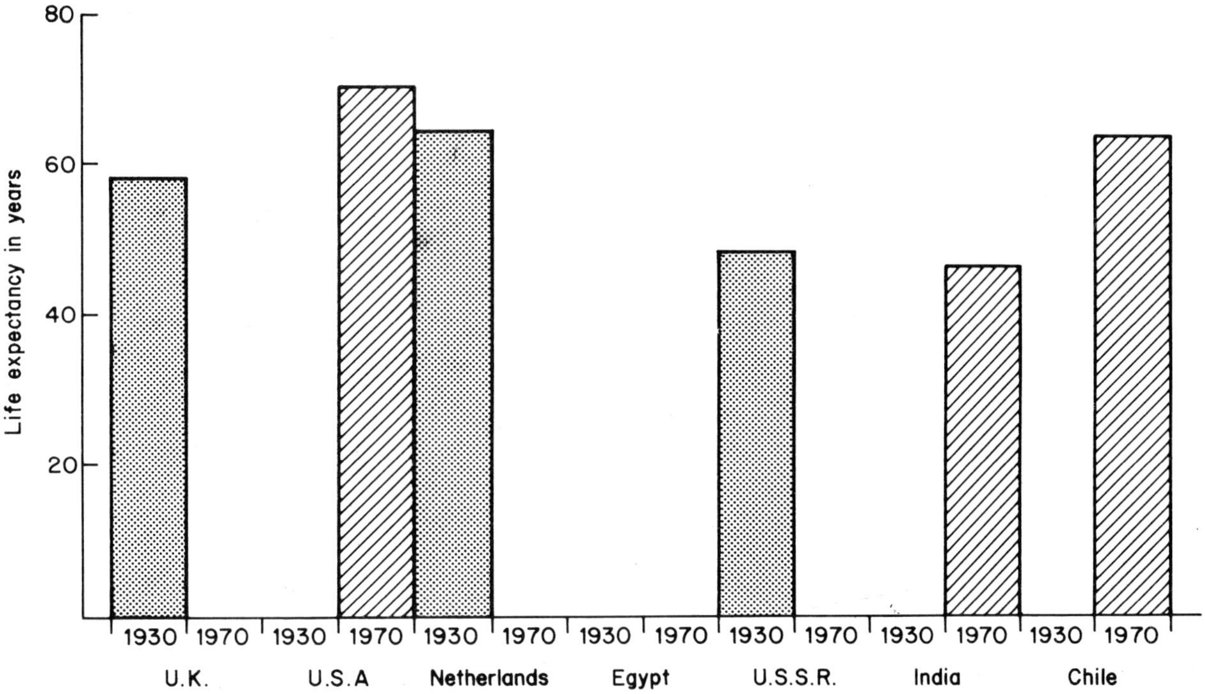

(ii) Which country showed the lowest life expectancy in 1930?

(iii) In how many countries has the life expectancy of people born in 1970 increased by less than twenty years since 1930?

(iv) Which country showed the greatest increase in life expectancy between 1930 and 1970?

(b) Three hundred years ago the average life expectancy in Britain was approximately 35 years.

State THREE reasons (biological or health) why the life expectancy is now double that figure.

(YR, Summer 1982)

The question places Biology in a world context with a comparative treatment which is not stated in the syllabus. The data and topic are such that additional questions requiring longer answers explaining why the life expectancies of the developing countries have not yet reached the same levels as those of the USA or UK could have been included, or requiring an analysis of the statistics to include completion of an average figure. Although this question is from a Human Biology examination, it would not be out of place in a Biology paper. If the future 16+ Biology syllabuses and examinations were to include a greater element of Human and Social Biology then it would be possible to place multicultural Biology as a central theme in the course.

Two CEE syllabuses from LR and EA were also examined (the latter is being run as a revised scheme for 1983). The EA syllabus in Applied Biology is divided into three sections, Microbiology, Ecology, and Social Biology. The Social Biology section covers many multicultural Biology topics, including sickle-cell anaemia, human blood groups, and their distribution, population control, the inheritance of disease, intelligence, inheritance and environment interaction, the validity of arguments about intelligence testing, human behaviour, health and hygiene, public health and mental health. Their content represents a significant advance on the sort of syllabus often offered at 16+. The multicultural content is also examined in a demanding manner with the depth and discriminative facility required for the proposed combined target group for single system 16+ examinations in Biology. The following questions illustrate this point and would not be out of place in some O-level short-answer papers.

> In human populations in tropical Africa, a defective gene occurs which, in the homozygous condition results in a fatal blood condition called *sickle-cell anaemia*. In this condition the red blood cells are sickle-shaped and produce defective haemoglobin which fails to carry enough oxygen round the body.
>
> Three genotypes and phenotypes occur in the population.
>
> HH - normal red blood cells - normal oxygen transport
> Hh - normal and sickled red blood cells - some anaemia but not fatal
> hh - sickled red blood cells - fatal anaemia in childhood
>
> A very important aspect of this condition is the effect that malaria (widespread in this region) has on people whose genotypes are HH and Hh. HH types suffer from a more severe form than Hh types, especially in childhood.
>
> Answer the following questions based upon the above information.
>
> (i) Name the biological process responsible for the origin of these defective genes.
>
> (ii) What kinds of red blood cells would be found in the Hh genotype?
>
> (iii) In the above example, does the H gene show complete dominance or co-dominance in the Hh condition?

B 10. Study the graphs in Fig. 3 before answering the questions which follow

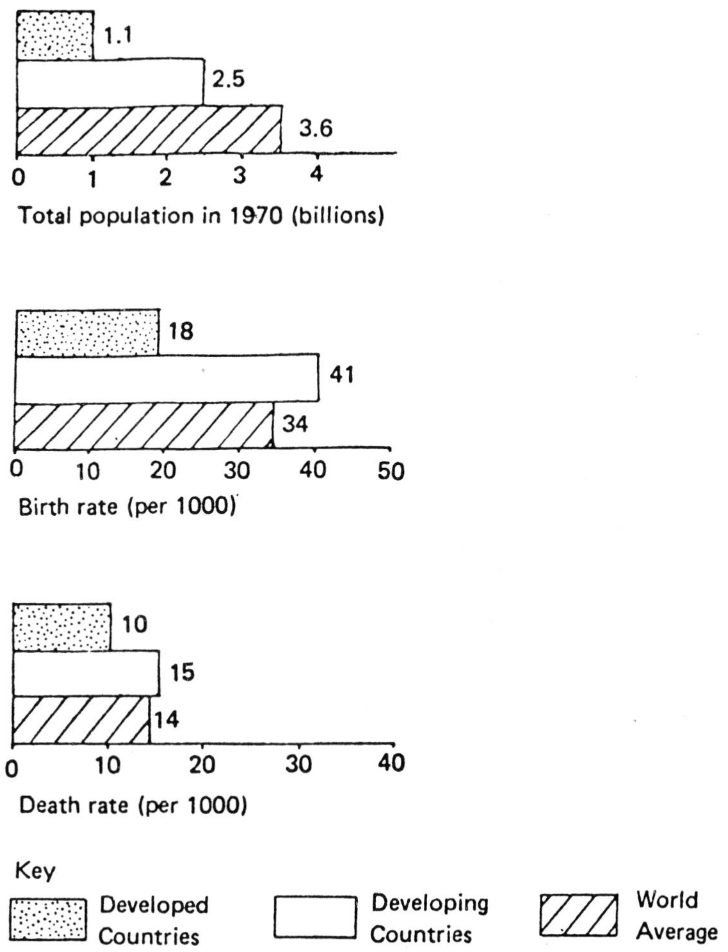

Fig. 3

(a) What is meant by the terms *developed* and *developing* countries?

(i) developed: ....

....

(ii) developing: ....

....

(b) Give two named examples of each kind of country.

Developed (i) (ii)

Developing (i) (ii)

(c) How much bigger is the population of the developing countries than the developed?

(d) What is the difference between the birth rate and death rate of the developing countries?

(e) Calculate the difference in answer (d) as a percentage difference of the birth rate.

(f) What important information can be gathered from the birth and death rate?

(g) Give two reasons why the birth rate of developed countries is lower than developing countries.

(EA, specimen question)

Although multicultural issues still appear rarely in 16+ Biology examination papers, the examples quoted here indicate that there is already a firm base on which to build future developments. Perhaps the situation is not quite as bleak as some teachers have painted it.

IV AGENDA FOR DISCUSSION

It is hoped that the ideas, criticisms, examples of good practice, and suggestions presented within the series might help stimulate the debate and serve as catalysts for future examination development. With this in mind, examining boards are urged to review their Biology examinations in the light of the following suggestions to see whether they meet the needs of all pupils in a multicultural society.

1 Interim reform of Biology examinations

Liaison with schools should continue in order to further the role of new examinations in terms of requirements for education in a multicultural society.

A project or coursework element should be made a compulsory component or interchangeable with a practical examination in all examinations (with weighting of no less than 20 per cent).

The examination boards should continue to review the questions set for cultural neutrality and bias, and should assess whether questions encourage candidates to display knowledge of the cultural and social aspects of Biology. Question-setting on these points would then stimulate teachers to devote more coverage to these neglected areas.

Examining boards should continue to consider the language used in questions and expected in answers.

Examining boards should make it known to schools that they already accept answers which use examples from all cultures if they provide acceptable alternatives to those familiar to the examiners.

2 Specific syllabus areas for inclusion

16+ Biology syllabuses should include the following topics:

Diversity and adaptation

The diversity of pigmentation, size, body hair, etc., and any relationship with climatic and environmental variations.

Examples used should include those from countries of origin of ethnic minorities as well as those found in the UK.

Animal nutrition and transport

Examples of foods rich in carbohydrates, lipids and proteins should include foods used by the cultural minorities.

A balanced diet should include diets found outside the UK, giving sources of protein, etc. in a comparative manner.

Energy requirements of individuals should be placed in relation to the availability of foods; malnutrition and starvation, and the reasons for their existence should be placed in the context of unequal resources.

Diet-deficiency diseases should be compared to obesity, and care taken to show that diet deficiencies can occur in all societies as products of poverty and wealth. Attention should be given to rickets as a culturally-related disease.

Diets with and without additives should be compared, and supplemented by illustration, with the occurrence of particular diet-related diseases such as bowel cancer, ulcers, etc. Links to demography should be included.

Study of world food supplies should include the problems of famine, unequal food distribution, farming resources and increasing population.

Dental hygiene should include a comparative view of the incidence of caries in developed and underdeveloped countries, and methods of oral hygiene should also be considered.

The ecosystem

The human being and the ecosystem should include the relationship of peoples to the environment, exploitation of resources, the unequal distribution of resources, the international use of pesticides and herbicides and the issues surrounding the type and location of such chemicals in the developing world, global conservation issues. Examples of ecosystems should include those from other countries.

Human blood groups: their possible origin and their variation should be considered as examples of differences related to racial groups.

Respiration

This should cover the adaptation of the human body to altitude and its significance. The link to thalassaemia should be included.

Reproduction

This should include a study of parental care in other societies, the use of contraception programmes (including depo provera and enforced sterilization as examples of controversial birth control), fertility and the onset of puberty related to factors such as race and diet, and differences in pre- and post-natal mortality between societies and racial/class groups.

Genetics

This should deal with sickle-cell anaemia and thalassaemia as examples of genetic diseases. There should also be scope for a treatment of the biological differences between races and their significance, including aspects of the IQ debate on race and intelligence.

Response

Study of the skin should include the significance of pigmentation, melanin, and carotene.

Disease

This should include diseases of world significance, vaccination programmes, disease control programmes, world health organizations, vectors of disease, diseases in UK.

3 Long-term suggestions

The examining boards in conjunction with the newly formed Secondary Examinations Council should establish a group to monitor the content of syllabuses and develop a more appropriate content based on the needs of the multicultural society.

Resources should be provided for the development of materials which can be used in the teaching of a multicultural Biology syllabus at 16+.

The effect and success of multicultural Biology should be monitored as part of normal programmes for review and revision of syllabuses and assessment

Consideration might be given in the new syllabus to specification of the key concepts and techniques/skills, with examples illustrating these for teachers and pupils.

4 Conclusion

One of the assumptions behind this report is the belief that all pupils should be able to learn and investigate the biological significance of their own and other groups' cultures. To do this, Biology examination syllabuses should encourage a feeling that all peoples have a contribution to make to the study of Biology, which should not be considered the property of the majority culture. Therefore all syllabuses should have a content which discusses the origins of all pupils and indicates the reasons for such variation within the human species. If this is achieved, then 16+ Biology will be capable of challenging racial and cultural stereotypes through both investigation and information. Such Biology syllabuses would encourage an atmosphere of cultural and racial tolerance.

REFERENCES

1 Schools Council, *Examining in a Multicultural Society*, report of conference held in 1981, 1982

2 Schools Council, *Education for a Multiracial Society: curriculum and context 5-13*, Schools Council Publications, 1981

3 Schools Council, *Education of Children from Ethnic Minority Groups*, Pamphlet 19, 1982

4 Department of Education and Science, *West Indian Children in our Schools. Interim Report of the Committee of Inquiry into the Education of Children from Ethnic Minority Groups* (chairman: Anthony Rampton, OBE) [Rampton Report], Cmnd 8273, HMSO, 1981

5 Afro-Caribbean Educational Resource Project, 'Follow Up to Rampton', personal communication, June 1982

6 R. Jeffcoate, *Positive Image*, Chameleon Books, 1979, p.37

7 Department of Education on Science Select Committee on Race Relations and Immigration, *Educational Disadvantage and the Needs of Immigrants*, Observations on the Report on Race Relations and Immigration, Cmnd 5720, HMSO, 1974

8 BBC TV, *South East at Six*, 14.2.83; *Tottenham Weekly Herald*, 28.2.83

9 Personal communication with Secretary of Haringey Black Pressure Group on Education

10 Van Rensberg, *Education for South Africa's Black, Coloured and Indian Peoples*, Erudition Books, 1973

11 Association for Science Education, *Alternatives for Science Education*, consultative document 1979, p.38

12 Ibid, p.24

13 H.E.R. Townsend and E. Brittan, *Multiracial education - need and innovation*, Schools Council Working Paper 50, Evans/Methuen Educational, 1973

14 Commission for Racial Equality, 'Schools and Ethnic Minorities' in *Comments on Education in Schools - a consultative document*, 1979, p.19

15 R. Willey, *Teaching in Multicultural Britain*, Schools Council Programme Pamphlet, Longman for Schools Council, 1982, p.27

16 Department of Education and Science, *Report by HMI on Aspects of Organisation and Curriculum in Seven Multi-Ethnic Comprehensive Schools*, DES, 1979, p.12

17 Association for Science Education, *Language in Science Education*, ASE, 1980

18 D. Barnes, T. Britton, H. Rosen and the L.A.T.E., *Language, the learner and the school*, Penguin, 1971

19 Department of Education and Science, *A language for life*, Report of the Committee of Inquiry appointed by the Secretary of State for Education and Science (chairman: Sir Alan Bullock, FBA) [Bullock Report], HMSO, 1975, para 20.5

20 Department of Education and Science, *Aspects of Secondary Education in England*, a survey by HM Inspectors of Schools, HMSO, 1979

21 *Bulldog*, issue 2, October 1977

22 Brent Teachers Association, *Multicultural Education in Brent Schools*, 1980, p.16

23 L. Claydon *et al.*, *Curriculum & Culture: Schooling in a Pluralist Society*, Allen and Unwin, 1977, p.59

24 A. Brues, *People and Races*, Macmillan, New York, 1977

25 S.R. Baker, *Race*, Oxford, 1974, pp.354-532

26 Ibid

27 G.O. Ferguson, 'The Psychology of the Negro' in *Archives of Psychology*, Vol. 5, No. 35, 1916, pp.1-138 and Baker (ref. 25), pp.472-478

28 J.A. Rogers, *Sex and Race*, Vol. 1, H.M. Rogers, St Petersburg, New York, 1940

29 A.J. Jensen, 'How much can we boost IQ and scholastic achievement', Harvard Educational Review, 39(1), 1969, pp.1-123

30 I. Gottesman, 'Biogenetics of race and class' in Katz and Jensen, *Social Class. Race and Psychological Development*, Holt, Rinehart and Winston (now Holt Saunders, Eastbourne), 1968, pp.1-11

31 P. Whitten and J. Kagan, 'Race and IQ' in *Readings in Physical Anthropology and Archaeology*, D.E. Hunter and P. Whitten (eds), Harper and Row, 1978, p.306

32 Ashley Montagu, *Statement on Race*, Schuman, New York, 1957, pp.1-110

33 Association for Science Education (ref. 11), p.38

34 Department of Education and Science (ref. 20), p.166

35 Jeffcoate (ref. 6), p.37

36 Department of Education and Science (ref. 20)

37 Department of Education and Science (ref. 20)

38 Community Relations Commission, *Education in a Multicultural Society - Syllabuses*, CRC, 1978, pp.35 ff.

39 'National Criteria for a Single System of Examining at 16+' (internal Council Paper 80/364 to Schools Council Examinations Committee), p.4

40 Schools Council, *National Criteria for a Single System of Examining at 16+*, Report from Science Committee's 16+ Sub-committee SC81/301, p.1

41 16+ GCE and CSE Boards' Joint Council for 16+ National Criteria for Biology, *Recommended 16+ National Criteria for Science*, 1983

42 Department of Education and Science (ref. 20)

43 Conference of Community Health Group for Ethnic Minorities, November 1982

44 Report of the Working Party on Education to the Prime Minister, January 1976, Trinidad and Tobago

Appendix A

THE GCE AND CSE EXAMINING BOARDS

The GCE Boards

Associated Examining Board (AEB)

Secretary: J. Day
Wellington House
Station Road
Aldershot
Hampshire GU11 1BQ

tel: 0252 25551

University of Cambridge Local Examinations Syndicate (Cambridge)

General Secretary: Dr F. Wild
Syndicate Buildings
17 Harvey Road
Cambridge CB1 2BU

tel: 0223 61111

Joint Matriculation Board (JMB)

Secretary: C. Vickerman
Manchester M15 6EU

tel: 061 273 2565

University of London School Examinations Department (London)

Secretary: A.R. Stephenson
66-72 Gower Street
London WC1E 6EE

tel: 01 636 8000

Oxford Delegacy of Local Examinations (Oxford)

Secretary: Miss C.G. Hunter
Oxford Delegacy of Local Examinations
Ewert Place
Summertown
Oxford OX2 7BZ

tel: 0865 54291

Oxford and Cambridge Schools Examination Board (O and C)

Secretary: A.R. Davis
Elsfield Way
Oxford OX2 8EP

tel: 0865 54421

Oxford and Cambridge Schools Examination Board (O and C)

Secretary: H.F. King
Brook House
10 Trumpington Street
Cambridge CB2 1QB

tel: 0223 64326

Southern Universities' Joint Board for School Examinations (SUJB)

Secretary: J. Symonds
Cotham Road
Bristol BS6 6DD

tel: 0272 36042

Welsh Joint Education Committee (WJEC)*

Secretary: J.L. Brace
Head of Examinations
Department: M. Jones
245 Western Avenue
Cardiff CF5 2YX

tel: 0222 561231

---

* CSE Board also.

The CSE Boards

Associated Lancashire Schools Examining Board (ALSEB)

Secretary: K. Tattersall
12 Harter Street
Manchester M60 7LH

tel: 061 228 0084/7

East Anglian Examinations Board (EA)

Secretary: Miss M.L. Thompson
The Lindens
Lexden Road
Colchester
Essex CO3 3RL

tel: 0206 49595

East Midland Regional Examinations Board (EM)

Secretary: D.J. Ramsden
Robins Wood House
Robins Wood Road
Aspley
Nottingham NG8 3NH

tel: 0602 295367

London Regional Examining Board (LR)

Secretary: D.H. Board
Lyon House
104 Wandsworth High Street
London SW18 4LF

tel: 01 870 3961

North Regional Examinations Board (NR)

Secretary: J.A. Winterbottom
Wheatfield Road
Westerhope
Newcastle upon Tyne NE5 5JZ

tel: 0632 862711

North-West Regional Examination Board (NW)

Secretary: J.E. Tipping
Orbit House
Albert Street
Eccles
Manchester M30 OWL

tel: 061 788 9521

South-East Regional Examinations Board (SE)

Secretary: P.N. Anderson
Beloe House
2-4 Mount Ephraim Road
Tunbridge Wells
Kent

tel: 0892 35311

Southern Regional Examinations Board (SR)

Secretary: H.G. Macintosh
Avondale House
33 Carlton Crescent
Southampton SO9 4YL

tel: 0703 32312

South-Western Examinations Board (SW)

Secretary: J. Edmundson
23-29 Marsh Street
Bristol BS1 4BP

tel: 0272 23434

West Midlands Examinations Board (WM)

Secretary: Dr G. Flack
Norfolk House
Smallbrook Queensway
Birmingham B5 4NJ

tel: 021 643 2081

West Yorkshire and Lindsey Regional Examining Board (WYL)*

Scarsdale House
136 Derbyshire Lane
Sheffield S8 8SE

tel: 0742 57436

Yorkshire Regional Examinations Board (YR)*

Secretary: B. Park
31-33 Springfield Avenue
Harrogate
Yorks HG1 2HW

tel: 0423 66991

---

* now merged to form the Yorkshire and Humberside Regional Examination Board

Appendix B

RESOURCES

Listed below are some publications, authors, and agencies which can be used to develop resources in Biology appropriate to the needs of a multicultural society. It is hoped that this will provide a stimulus for boards and teachers to appraise their usefulness, and to develop their own materials.

Texts

John Baker, *Race*, Oxford University Press, 1974. Definitive book on racial evolution, origins, meaning of, trends, colour, odour, physical differences.

D. Baldwin, *Then and Now*, Child Development series, Longman, 1978.

Fraser Brockington, *World Health*, Pelican, 1958 (out of print). A very informative but dated book.

C. Clarke, *Human Genetics and Medicine*, Studies in Biology series No. 20, Edward Arnold, 1977.

A.G. Clegg and P.C. Clegg, *Man against Disease*, Heinemann Educational, 1973. Good background information and detail.

L.B. Cook, *Caribbean Agricultural Science* Bks 1 and 2, Collins, 1978. Good section on economic crops of the Caribbean, their cultivation, diseases, use.

T. Geoffrey-Taylor, *Principles of Human Nutrition*, Studies in Biology series No. 94, Edward Arnold, 1978.

A. Godman and C. Webb, *Certificate Human and Social Biology*, Longman, 1974. Gives a simple treatment of the relation of Vitamin D synthesis to sunlight.

J.J. Head, *Science Through Biology*, Edward Arnold, 1976. Reasonable section on sickle-cell anaemia.

Health Education Council, *Health Education Materials for Ethnic Minorities*, 1984. A source list which is regularly updated. Recommended.

D.G. Mackean, *Introduction to Biology - tropical edition*, John Murray, 1977. Excellent section on food storage plants, e.g., yams, cassava, their growth and cultivation.

Van Overbeck, *Lore of Living Plants*, National Science Teachers' Association, Washington, 1964. Contains a complete section on nutrition science and the individual.

R. Price, *Beginning Biology*, Cambridge University Press, 1983. A global perspective with good information for question-setting.

R. Taylor, *World Health*, Macdonald Educational Colour Units, 1977.

T. Williams, *Drugs from Plants*, Sigma Books, 1947. Excellent specialized book. Includes chapter on the contribution of Egypt to modern medicine.

Agencies and individuals

Community Health Group for Ethnic Minorities, 28 Churchfield Road, London, W3 6EB. Wide range of information available on diet, morbidity, mortality, race, hypertension, mental health and race; rickets, women's health available from group serviced by GPs, pharmacists and nurses.

Ealing, Hammersmith and Hounslow Area Health Authority's health education service has compiled a list of resources including films and speakers on many aspects of minority ethnic health facilities in Asian, Greek, and Turkish languages.

Organization for Sickle Cell Anaemia Research, Wood Green High Road, London N22.

G. Khontiorghes, researcher on thalassaemia, University College Hospital, Gower Street, London W1.

Third World Science Project, Professor Iolo Wyn Williams, School of Education, University College of North Wales, Bangor, Gwynned. The project has produced a series of readers, including practical work, for UK teachers to enrich school science and their view of science in the developing world. There are units on salt, soap, housing, carrying loads, fermentation, clay pots, dental care, distillation, vegetable oils, methane digestors, iron smelting and useful plants.

Appendix C

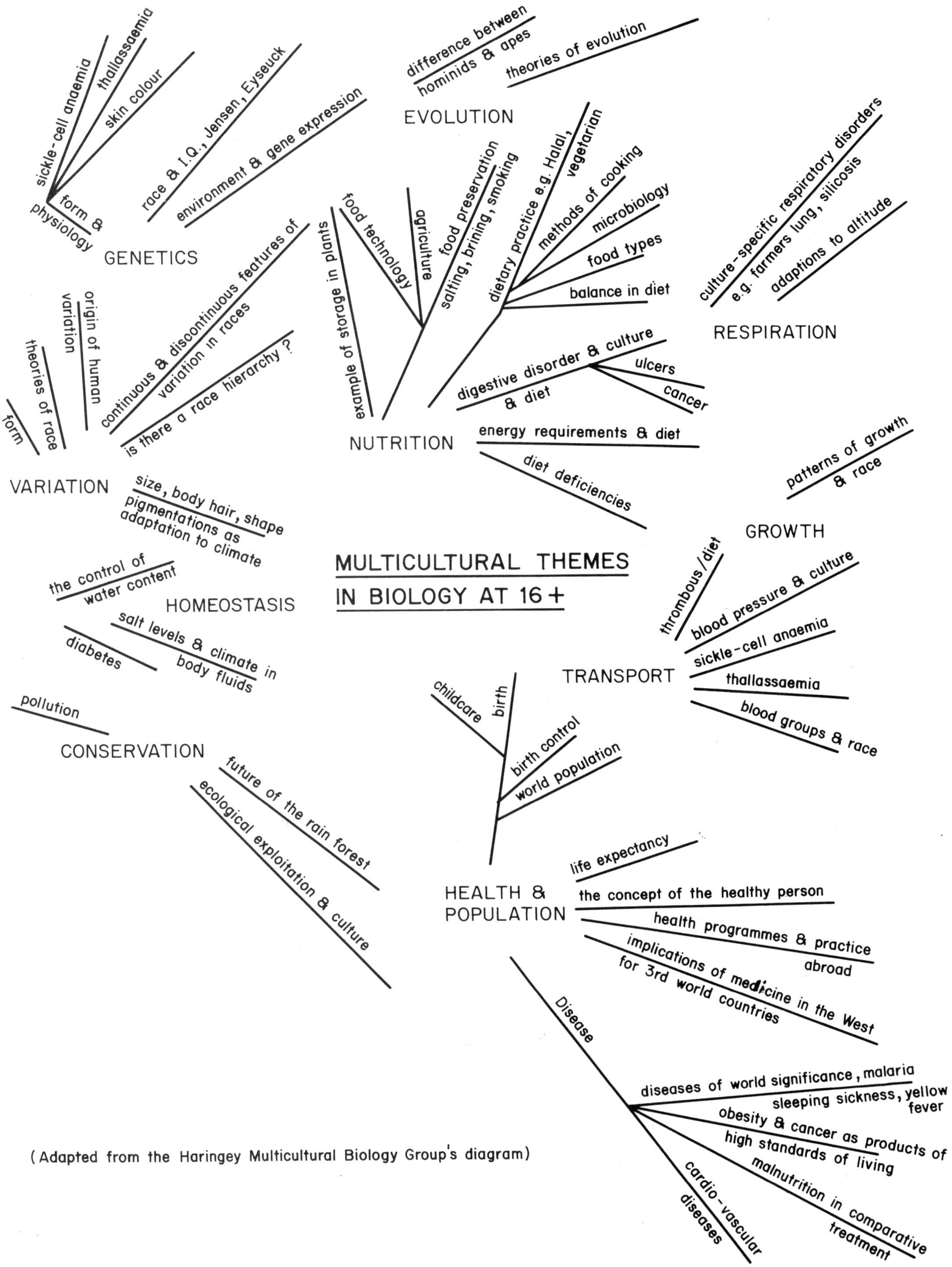

(Adapted from the Haringey Multicultural Biology Group's diagram)

PROGRAMME 5 MONITORING AND REVIEW GROUP

A.H. Jennings (Chairman)	Formerly Headmaster, Ecclesfield School, Sheffield; acting Chairman, Schools Council (Secondary Heads Association)
R. Aitken	Director of Education, Coventry (Association of Metropolitan Authorities)
SI B.C. Arthur	HM Inspectorate of Schools
J.J. Billington	Deputy Headmaster, High Pavement Sixth Form College, Nottingham (Assistant Masters and Mistresses Association)
D. Foster (to December 1981)	Headmaster, Churchill School, Bristol (CSE examining boards)
W.S. Frearson (from March 1982)	Ashford College, Middlesex (CSE examining boards)
J.C. Hedger	Department of Education and Science
P.M. Herbert	The Elliott School, London SW15 (National Association of School-masters/Union of Women Teachers)
H.F. King (from April 1982)	Secretary, Oxford and Cambridge Schools Examination Board (GCE examining boards)
D.I. Morgan	Vice-principal, W.R. Tuson College, Preston, Lancs (National Union of Teachers)
R. Potts	Harraby Comprehensive School, Carlisle (National Union of Teachers)
A.R. Stephenson (to March 1982)	Secretary, University of London School Examinations Department (GCE examining boards)

PROJECT STEERING GROUP AND PROFESSIONAL SUPPORT

Project steering group

F. Cummins (Chairman)	Headteacher, Thomas Telford High School, Sandwell	*representing CSE examining boards*
H. Kelly (from January 1983)	Secretary, Association of LEA Advisory Officers for Multicultural Education	*representing LEAs*
E. Parsons	Deputy Headteacher, Whitley Abbey School, Coventry	*representing teachers*
C. Power (to September 1982)	Multicultural Adviser, London Borough of Haringey	*representing LEAs*
A. Shaffer	Lecturer in Sociology, Kingsway Princeton College, London WC1	*representing GCE examining boards*
P. Singh	HM Inspectorate of Schools	

Schools Council

G. Bardell (Lead officer)	Assessment and Examinations Team	
A. Craft	Coordinator for Multicultural Education	
G. Insull (from August 1982)	Administrative support	
S. Phillips	Lecturer in German, Cross Keys Tertiary College, Gwent	*representing Modern Languages Committee*
P. Scott (to July 1982)	Administrative support	
B. Thompson	Science teacher, St Paul's RC School, Leicester	*representing Working Party on Multicultural Education*

Professional support

Community language consultants

M. Roussou	Greek Coordinator Schools Council Mother Tongue Project
R. Russell	Emeritus Reader in Urdu School of Oriental and African Studies, University of London

External readers

B. Allison	Professor of Art Education, Leicester Polytechnic	*Art and design*
R. Bruce	Secondary Adviser in Religious Education, Inner London Education Authority	*Religious studies*
P. Burdett	Thornes House School Thornes Park, Wakefield	*Biology*
M. Christodoulides	Assistant Examiner, School Examinations Department, University of London	*Community languages: modern Greek*
E. Gray	Head of English, Wakefield Girls' High School	*English*
Dr R. McGregor	Lecturer in Hindi, Faculty of Oriental Studies, University of Cambridge	*Community languages: Hindi*
R. Russell	Emeritus Reader in Urdu School of Oriental and African Studies, University of London	*Community languages: Urdu*
I. Shelton	Deputy Headteacher, Sale Girls' Grammar School, Trafford	*Social sciences*
Dr B. Tolley	Lecturer in Education, University of Nottingham	*History*
M. Walshe	Staff Inspector, Inner London Education Authority	*Home economics*